Uncomfortable Situations

Uncomfortable Situations

Emotion between Science and the Humanities

DANIEL M. GROSS

The University of Chicago Press
Chicago and London

The University of Chicago Press, Chicago 60637
The University of Chicago Press, Ltd., London

Published 2017
Printed in the United States of America

26 25 24 23 22 21 20 19 18 17 1 2 3 4 5

ISBN-13: 978-0-226-48503-4 (cloth)
ISBN-13: 978-0-226-48517-1 (e-book)
DOI: 10.7208/chicago/9780226485171.001.0001

Library of Congress Cataloging-in-Publication Data

Names: Gross, Daniel M., 1965– author. | Preston, Stephanie D. (Stephanie Delphine)
Title: Uncomfortable situations : emotion between science and the humanities / Daniel M. Gross.
Description: Chicago ; London : The University of Chicago Press, 2017. | Includes bibliographical references and index.
Identifiers: LCCN 2016055406 | ISBN 9780226485034 (cloth : alk. paper) | ISBN 9780226485171 (e-book)
Subjects: LCSH: Emotions in literature. | English literature—19th century—History and criticism. | English literature—18th century—History and criticism. | Darwin, Charles, 1809–1882. Expression of the emotions in man and animals. | Equiano, Olaudah, 1745–1797. Interesting narrative of the life of Olaudah Equiano. | Radcliffe, Ann Ward, 1764–1823. Romance of the forest. | Austen, Jane, 1775–1817. Sense and sensibility. | Emotions (Philosophy). | Psychology and literature. | Neurosciences and the humanities.
Classification: LCC PR149.E55 G76 2017 | DDC 820.9/353—dc23 LC record available at https://lccn.loc.gov/2016055406

♾ This paper meets the requirements of ANSI/NISO Z39.48-1992 (Permanence of Paper).

Contents

Acknowledgments

As a sequel of sorts to *The Secret History of Emotion*, this book is particularly indebted to a series of conversations that advanced my thinking substantially. Darwin material became the foothold for this book after my UC Irvine colleague, Jonathan Alexander, asked me to develop something surprising for a local symposium on work across the sciences and the humanities. Subsequently the Darwin piece was published in *Critical Inquiry* 37, no.1 (2010) and the book project launched, but only after thoughtful feedback from cross-disciplinary gatherings at the University of Waterloo, the University of Pittsburgh, and the University of Manitoba. Lauren Berlant, who edited my Darwin article, has been instrumental as I developed this project in a world now familiar with Emotion Studies thanks, in part, to her efforts.

My chapter on Gibson and Gothic environments shaped up in collaboration with my UCI colleagues Julia Lupton and Jayne Lewis, both of whom read heaps of preliminary material that I designed with their sensibilities in mind. I thank participants in my graduate seminars Cognitive Approaches to Literature (2012) and Affect Criticism (2016), who helped me figure out how the book project might arrive on the scene. Particular insights are due to a few of these individuals: Brendan Shapiro alerted me to the work of David Herman, Maureen Fitzsimmons reminded me about Kahneman's relevance to literature, Jens Lloyd paid special attention to environmental details in Radcliffe and Austen, and Elizabeth Mathews got me thinking about "emotional taste" and its implications for literary history. Audiences at the Modern Language Association Conference, the National Communication Association Conference, and Case Western Reserve University were particularly helpful as I developed ideas in this portion of the book.

My chapter on mixed feelings in Austen benefited from conversations around visits to UCLA and Northwestern University, and from the remarkable cross-disciplinary UC Irvine Persuasion Conference 2016, also organized by Jonathan Alexander.

Not surprisingly the epilogue of this book is indebted first to coauthor Stephanie Preston, who has served as a challenging interlocutor periodically, going back to our days at the University of Iowa when we met in 2003 preparing for our conference The Promise of Empathy. An earlier version of this chapter was published in a book I coedited with Frank Biess, *Science and Emotion after 1945: A Transatlantic Perspective.* Frank was a catalyst as I worked out the methodology for *Uncomfortable Situations* in response to the UC San Diego History of Emotions conference he organized. Also at the conference I met William Reddy, who has subsequently shaped my work as a formidable fellow traveler in the subfield.

Broad support and close reading was provided by Thomas Rickert, Steve Mailloux, Jami Bartlett, Jim Steintrager, and Irene Tucker. Libby Catchings taught me a great deal about prosthetic emotion, and she provided expert research assistance as a number of images were wrangled into place. Thanks also to Will Jones for his work on the recalcitrant McDonald's Happy Meal image. At the University of Chicago Press, Douglas Mitchell continues to provide an extraordinary venue for publications in rhetorical studies. Finally, thanks to my family for their ongoing interest in, and forbearance toward, my academic work. It should go without saying that this project in Emotion Studies would not be the same without the people who surround me. For that I am grateful.

INTRODUCTION

Uncomfortable Situations

Imagine this scenario that comes from the cognitive science literature:

> You're at a dinner party with friends. A debate about a contentious issue arises that gets everyone at the table talking. You alone bravely defend the unpopular view. Your comments are met with sudden uncomfortable silence. Your friends are looking down at their plates, avoiding eye contact with you. You feel your chest tighten.[1]

Despite myself and what should be the numbing familiarity that comes with rereading, this passage does make me feel uncomfortable. In this situation would I be angry at myself for going out on a limb, or fearful that my friends might distance themselves further? I don't know. I guess it depends upon the situation that would need more Austenesque details about my relationship to interested onlookers and the view I defended. In fact it's not hard to imagine a scenario where along with isolation fears, and anger at myself, I feel pride for defending the unpopular view to the point where I discover that my friends aren't friends at all . . . now also indignation and perhaps sadness. This kind of event, in other words, need not unfold with me feeling mixed in the midst of a stable situation; the event could find me resolute, with those around me appearing one way or another. Then how do we understand the feelings that were supposed to be friendly? Basic emotions slip away as we stumble over voice: appearing to whom? Are feelings prior to the event somehow unfelt retroactively, just as Lady Russell in Austen's *Persuasion* must "learn to feel that she had been mistaken" about Anne's suitors Elliot and Wentworth?[2] Though melodramatic in this opening instance, such discomfort in the midst of a dynamic situation is familiar, even commonplace.

Another important complication lies at the heart of this opening scenario from the cognitive science literature. Again with Austen, it's worth pointing out that my mixed feelings would be determined not by any predictable calculation internal to the scene as we imagine it arm's-length on the page, or on stage, or on screen. Mixed feelings are determined with respect to audiences who might be outside the immediate situation, including you. Whereas Lauren Berlant defines a situation as "the state of things in which *something* that will perhaps matter is unfolding amidst the usual activity of life," pointing in *Cruel Optimism* to the situation comedy or the police procedural—"We have a situation here"[3]—I will instead emphasize in this book how a situation is characterized by emotional instability defined rhetorically, which is how you and I appear on the scene. Methodologically, we can't get at the uncomfortable situation adequately without a rhetoric of audience investments, as I will illustrate over the course of this study; and along the way it should become evident how this approach differs from its closest neighbors (like Sigmund Freud on psychoanalytic ambivalence, Erving Goffman on sociological uneasiness, and Martin Heidegger on the philosophical bad mood, *Verstimmung*, when everyday concerns go awry). It is also why sentimental literature, not television, provides for my study the optimal research domain: the genre was characterized by its rhetoric of emotional instability, as I will explain focusing on an example from Sterne.[4] Sentimental literature around the turn of the eighteenth century shows in appropriate detail how—despite the optimism analyzed by Berlant, and which I explore across a handful of disciplines—situations are basically uncomfortable.

Or to put this another way—and responding to James Chandler's important work on the topic, sentimental literature establishes the norms for well-being we still reference when gauging an emotional reaction like those above.[5] The way Chandler tells the story, Adam Smith makes the key move when he stipulates that sympathy is not feeling what another person feels, but rather feeling what we ourselves *should feel* in a like situation. This philosophical ethics in turn makes room for the "sentimental moment in the history of fiction" (302) when Laurence Sterne plays a crucial role, as he develops novelistic techniques for triangulating emotional situations so that immediate reactions, or naïve sentimentalism, have their moment, but always invite their qualification or indeed their very undoing when the "syntax" of the sentimental moment is revealed—a move mastered later by Blake (277).[6] Chandler then argues that this perspective-taking on emotional situations is precisely what defines the sentimental mode that persists through its crucial reinvention by Frank Capra's classic cinema where America's intimate public sphere takes new form: "Adam Smith Goes to Hollywood" (13).

Chandler tells a compelling origin story. But his ultimate focus on the spectator guides his analysis in a different direction, away from the situation per se, where his spectator is supposed to feel one way or another. Instead I posit the sentimental moment in the history of fiction initiates what we now understand as our basic situation, which is supposed to be comfortable if we follow key literature across the disciplines, but appears anything but if we return to the formative moment and see how the terms of our comfort take shape polemically. So instead of an archaeology of sympathy, one might call this a critical history of well-being, where sentimental literature of the eighteenth century provides the pivot point.

At this point I should explain how this book has emerged out of my earlier critique of emotion science that left little room for the cross-disciplinary conversation I now find essential. In 2006 I offered *The Secret History of Emotion: From Aristotle's* Rhetoric *to Modern Brain Science* as a contribution to a humanities, and more exactly a rhetorical critique of emotion science as it had emerged powerfully in the academic and popular presses. My objective was three-fold. First, I wanted to demonstrate how the new brain science of emotion doesn't account for basic phenomena that are situated beyond individuals, for instance the emotional response to Princess Diana's death lost on a suffering nobody. Second, I wanted to show how the tradition after Aristotle's *Rhetoric* offers alternatives for understanding emotions as social phenomena, which is something that leading humanists like Martha Nussbaum and Richard Sorabji should have remembered as they rushed toward the latest brain science. Third, I tried to show how the foundations of modern brain science emerged in early modern Europe precisely at the expense of this rhetorical tradition that continues to linger suggestively in our background.

Unsurprisingly, I was then challenged by simple questions: "Are you anti-science?" and "Can't there be a worthwhile science of emotion?" In 2006, it did seem a polemic was called for in the face of overwhelming odds. There was little room for qualification and common ground because my critique might then wind up looking like science-humanities "consilience" along the lines of E. O. Wilson and Steven Pinker, which makes room for the humanities only insofar as they are dependent upon scientific inquiry. Passions needed stirring; some outrage woven through the genealogy seemed the only way to break through, especially to academic humanists who were making room for scientific programs that could overwhelm. Nevertheless, this legitimate challenge stuck with me, and since then I have been working on a response. Now both within and without brain science, critiques complementing mine are more firmly established as I'll demonstrate, which makes new room for cross-disciplinary work of a different sort: I'm referring most prominently to a "situated emotion" critique of the basic emotions program mentioned above, a "situated cognition" critique of computational psychology, and a critique of evolutionary psychology from many angles, including cognitive scientific. So, yes, I now agree, from the perspective of uncompromised humanities inquiry, there is a worthwhile science of emotion.

In their 2009 *Cambridge Handbook of Situated Cognition* chapter titled "Emotions in the Wild," philosophers of science Paul Griffiths and Andrea Scarantino start to map, by way of omission, the potential contribution of the humanities to a science of emotion.[7] Drawing heavily on transactional accounts of emotion proposed by some contemporary psychologists, Griffiths and Scarantino shift our theoretical focus from terrible snakes and bawling babies to what they consider neglected phenomena like a wedding ceremony, where the "cultural scaffolding of emotional performances" (445) obviously exceeds individual feelings and instead supports the right emotions at the right times necessary for society to "function smoothly." Among other factors, they are interested in the transactional character of emotion (sulking to get a better deal in a relationship, 440), material factors, including emotional capital (resources associated with having a specific social status, gender, etc., 444), emotional affordances (the wedding ceremony itself, the venue . . .), and emotional indeterminacy in the art of misleading others and precipitating context-dependent disambiguation—a capacity without which human communication and most wedding ceremonies would look very different indeed.

But when we want to model the cognitive dynamics of emotion ecologically, narrative literature, including fiction, has a special appeal beyond the social sciences. Take love, to continue the example beyond sentimental literature, in the context of marriage. For example, the *Much Ado about Nothing* wedding ceremony of Beatrice and Benedick is perversely shaped by the insinuation that these sometimes rivals were too proud to love one another. Could this emotional situation be characterized adequately without the attending fiction that shows how love emerges in this case from a contrary impulse? Although cultural anthropology can rival narrative literature as thick description, it is obligated to what Clifford Geertz calls "concrete social events and occasions"[8] to a degree that precludes ecocriticism of the rhetorical sort I'll practice in chapter 4, referencing James J. Gibson. Or turning to a foundational work of sentimental fiction, where I spend most of my time in this project, consider the double wedding ceremony in Sarah Fielding's 1744 novel, *The Adventures of David Simple: Containing an Account of His Travels through the Cities of London and Westminster, in Search of a Real Friend.*[9] In this scene, risqué French romances provide some of what Griffiths and

Scarantino call the "cultural scaffolding of emotional performances" against which Fielding writes, when she explicitly refuses to describe the beauties of nature that adorn the heroes and heroines, choosing instead to reacquaint readers with the world-contrary "minds" of the key characters, according to which their oddly benevolent schemes of life might be followed (236–238). Where in the *Cambridge Handbook* Lawrence Barsalou admits, "the situations that pervade cognition are open ended and difficult to enumerate" (256), I agree, suggesting more optimistically that literature including nonfiction like Equiano's *Interesting Narrative,* and then differently the sentimental novel, can help us understand how situations take shape. Let me illustrate by way of a thought experiment, otherwise known as a very modest work of narrative fiction again focused on love.

An alien visitor after the apocalypse wants to figure out what humans meant by "romantic love." Perhaps something might be learned about love by putting a couple of ratty survivors in a room together and watching them for a long while. But perhaps not, or not much. Alternatively the alien might read Steven Pinker on the topic but wouldn't learn much beyond what supposedly happened over 10,000 years ago, if anything romantic happened at all.[10] A good historical dictionary definition would help, though the required abstraction would also distance the alien; remember that OED definitions build up inductively by way of direct quotations from the literature that provides the context for interpretation. In fact our alien, like the historian Lawrence Stone, could do worse than to learn about romantic love by reading a bunch of romances and novels (i.e., beyond the early eighteenth-century sense of French "romance" mentioned by Fielding). And why is that? Because romances and novels in Stone's largely eighteenth-century European repertoire show how romantic love works ecologically, including for Stone love at first sight (what might now be called a cognitive-perceptual phenomenon), love above all other considerations including material (i.e., love as a cognitive evaluation), and love as an admirable personal emotion given full rein "no matter how exaggerated and absurd the resulting conduct may appear to others."[11] Let's call this love a marginally acceptable antisocial stance that may be head over heels, or upside down. Second, romances and novels provide a concrete sense for the historicity of romantic love. "No doubt," qualifies Stone, certain young people have always defied the conventional wisdom that condemns such mental disturbances and have fallen head over heels in love (282). But it wasn't until the later European eighteenth century that Stone sees the growth of marriage for love—instead of marriage for interest—as a respectable motive among the propertied classes, accompanied by a rising flood of novels devoted to the same theme (284). And though Stone's argument was subject to vigorous criticism, which has typically relied upon counterevidence that would qualify his more dramatic claims about historical change,[12] Stone's basic methodology that includes the study of literature remains very appealing to anyone—including aliens or us—interested in the history of emotion.

More recently in *The Making of Romantic Love: Longing and Sexuality in Europe, South Asia & Japan, 900–1200 CE*, leading historian of emotion William M. Reddy has refined a methodology that includes a wide range of literary artifacts as they speak to the relevant sciences of emotion, and he dates the phenomenon earlier because he is less interested in romantic love as the emotional cornerstone of a new social order with companionate marriage at its center, and more interested in romantic love appearing as a specifically Western and Western-influenced alternative to desire-as-appetite, especially as this model of desire was canonized in the Gregorian Reform of approximately 1050–1200 (17, 26).[13] Like Stone (and our alien), Reddy looks for his evidence amongst "love literatures and documents" (9); but unlike Stone, Reddy references the latest cognitive science and psychology as he explains why literature is relevant beyond what one might find in the brain-focused laboratory. An emotion like romantic love, according to Reddy, is a phenomenon instantiated by its expression as it "enhances the background activation level of that range of thought material that is the emotion" (8); "I love you" as an *emotive* does not merely describe some preexisting condition, it instantiates the emotion and it does so unpredictably because the outcome (and I would add given the following example from Sterne, the intention) is not certain. So for Reddy, love literatures provide essential materials for a study on romantic love because that is precisely where an emotive like the declaration of love appears through its "material": its generic and social viability, its concrete opportunity, its deployment, its precise form, its receptions. At this moment cognitive behavioral therapy appears for Reddy as he tries to explain how such an emotion is experienced. But his point about emotional unpredictability and the "situated-ness" (27) of sexual desire establish what sort of brain science won't help in this instance and others like it. "There is, therefore, nothing in the latest neuroscience research on sexual desire, sexual arousal, or romantic love that permits one to conclude these states are caused or orchestrated by hard-wired brain systems. Instead, the most sophisticated methods presently available yield results that are compatible with the idea of a substantial role for cultural determinants in the occurrence of and the experience of such states. This conclusion is consistent with trends in many other areas of cognitive neuroscience . . ." (15). Complementing the work of a historian of emotion like Reddy, my perspective throughout this project is that a specific kind of cognitive science research—the situated variety—provides the best point of contact for humanities work on emotion.

A Sentimental Journey

Now, my central examples from Sterne. As the very title suggests, Laurence Sterne's *A Sentimental Journey through France and Italy* (1768) is a masterful study of emotional situations of a certain sort, namely sentimental. The best way to see how this works, not surprisingly, is to read one of the emotional situations we find there, and characteristically, few situations in Sterne go without their own reflections built-in, without "commentary upon the text," as Yorick will actually say in the passage itself. The setting: protagonist and Sterne alter-ego Yorick, having just arrived in Calais on his way to Paris, finds himself face-to-face with a lady he had noticed earlier, and he takes her hand:

> THIS certainly, fair lady! said I, raising her hand up a little lightly as I began, must be one of Fortune's whimsical doings: to take two utter strangers by their hands—of different sexes, and perhaps from different corners of the globe, and in one moment place them together in such a cordial situation as Friendship herself could scarce have achieved for them, had she projected it for a month—
>
> —And your reflection upon it, shows how much, Monsieur, she has embarrassed you by the adventure.—
>
> When the situation is what we would wish, nothing is so ill-timed as to hint at the circumstances which make it so. You thank Fortune, continued she—you had reason—the heart knew it, and was satisfied; and who but an English philosopher would have sent notices of it to the brain to reverse the judgment?
>
> In saying this she disengaged her hand with a look which I thought a sufficient commentary upon the text.
>
> It is a miserable picture which I am going to give of the weakness of my heart, by owning that it suffered a pain, which worthier occasions could not have inflicted.—I was mortified with the loss of her hand, and the manner in which I had lost it carried neither oil nor wine to the wound: I never felt the pain of a sheepish inferiority so miserably in my life.
>
> The triumphs of a true feminine heart are short upon these discomfitures. In a very few seconds she laid her hand upon the cuff of my coat, in order to finish her reply; so some way or other, God knows how, I regained my situation.[14]

Just so we are clear. Sterne gives us the following situation: 1) Yorick and the lady are in a situation together as he grabs her hand; 2) Yorick reflects on that "fortunate" situation noting how they are given to each other appropriately: similarly classed we now know as potential friends, and oppositely sexed—how romantic. 3) The lady remarks upon his reflection that ruins the moment as it draws attention to itself—recalling those English philosophers who do so awkwardly[15]—and she drops his hand immediately. 4) Time is then taken to call the situation uncomfortable—soon to be relieved by the lady said to be short upon such "discomfiture"—and Yorick claims the situation regained by way of the lady's hand but ultimately secured somewhere else, "God knows how." So an outside perspective *is* figured in this passage as "worthier occasions" for suffering make a Smith-like appearance; but instead of Smith, we get all sorts of admitted limitations, set within the frame of a colloquial God whose calculations are immeasurable.

What defines the "situation" in this case? And what would it mean for the situation to be "sentimental"? In fact, "to think of making love by *sentiments*!" seems to Yorick appalling, or at least this is how he later represents it to the lady before she concludes "you have been making love to me all this while" (26–27). In other words, yes and no. *No*, the situation is not sentimental because it could never be that thing once attention is drawn, which it must be from the outset because this is a piece of writing that does nothing but draw attention—that's what literature does, after all. The keyword is in the title. But *yes*, the situation is sentimental because the word is in the title, and what could be more sentimental than this situation, which must in fact be something in the first place, if it can be ruined? When it comes to sentiment who indeed needs the brainy English philosophers or, I would add, the brain scientists, when we have Sterne (he seems to say)? In a passage like this, Sterne shows us how sentiments work, and he makes it clear this is exactly what he is doing as the thing itself—true sentiment—appears by way of its negation and thus we are left with only the places where we expect it to be generically. That's what defines the situation in this case: a set of romantic expectations given appropriately with class and opposite sex at the core, and then destroyed by drawing attention to these mundane facts, which leaves the characters at a loss, and the readers at a loss, as we are somewhat relieved—comforted—precisely by the expectations that cannot be met. Who, after all, would find it romantic if sentiments were clear and they added up just like they were supposed to (also a concern for the post-utilitarian Austen we will see)? And who would want to read such a story, let alone lead such a life? Certainly not Sterne himself, who through this work comes to define the sentimental novel and the very phenomenon—sentimentality—like nobody else besides perhaps Rousseau in France and Goethe in Germany. Alternatively, we can say that Sterne regularly disappoints expectation at the level of character, and that is how he moves the scene forward; at the same time Sterne disappoints expectation at the level of the reader ("Where's the sentiment?") and that is how the reader is moved. Utter predictability would never work. Negation both at the level of semantic values and theology are crucial for Sterne's rhetoric of emotion, as I pursue in chapter 1.

Now after Sterne we can return to the beginning with some notes on methodology—specifically what my home field in the humanities brings to bear. A sign of the times, our opening thought experiment and the initial anger-or-fear alternative is plucked out of an important 2011 *Neuropsychologia* study, "Grounding Emotion in Situated Conceptualization," published by senior researchers Lawrence Barsalou, Lisa Feldman Barrett, and colleagues. In that study the researchers set out to critique and provide laboratory evidence refuting the "basic emotion program" attributed prominently to Paul Ekman, replacing it with a model where emotion is situated. (On critical trials during brain scanning, participants rate physical danger and "social evaluation" situations like the one above.) The effort is important because it decouples emotion from narrow brain science and evolutionary psychology, while introducing into the laboratory some noteworthy context variables, including audience. At the same time the effort is hampered, I'll argue in chapter 4 and with Stephanie Preston in the epilogue of this book, because it does not take humanities-based inquiry seriously enough, which leads to some troubling loose ends we glimpse above. How does one best characterize a laboratory study of emotion in terms of 1) its exigencies like social improvement, where for instance more empathy is pursued; 2) its key terms like *anger* as they impinge upon ordinary language and folk psychology; 3) its mediation like verbal and visual depiction; and 4) its interpretation, where causal explanations for the observed phenomena are regularly ventured, and where rhetorical figuration (like Sterne's negation) directs one beyond the laboratory? Because emotions cannot be isolated in the laboratory to the point where these context variables are completely neutralized—not without losing the phenomenon at hand, for example, anger or fear or empathy—a contribution from the humanities suggests itself strongly.

The Rhetoric of Emotion

Let me be clear. Debating "Emotions: hardwired or socially constructed?" is distracting because it collapses the phenomenology that is where we in fact spend most of our time. So to replace this binary debate I offer the following phenomenology of emotion, which takes shape around certain types of emotional transactions. Hence we can also call this more exactly a "rhetoric of emotion" because we are both describing transactions along the lines of Griffiths and Scarantino, and *asserting relationships* when we take them up in a certain way. Those relationships can be, for instance, as follows:

1) *Mimetic.* In literary criticism what Sianne Ngai summarizes as the "literary strategy of sympathy," or a "perfectly symmetrical circuit of affective 'communication' in which the reader feels what a character feels."[16] Tears produce tears, terror produces terror. Analog models in the natural sciences include mirror-neuron theory, and some aspects of the theory of how we recognize emotion in the face: if we see that someone else looks terrified we might feel terrified sympathetically, and for reasons that are biologically adaptive because we might very well be terrified of the same thing. Or adaptively once again, we are aligned to the emotions of our close kin like my baby, where there is often good reason to feel anger, disgust, fear, happiness, sadness, and surprise—sympathetically.

2) *Cathartic.* Witnessing emotional situation X can regularly produce in us a very different emotion, of a particular sort Y. Consider famously Aristotle's *Poetics* where the suffering of Oedipus is supposed to produce a sense of pleasant relief in the reader or spectator.

3) *Projective.* Here we have, in Ngai's helpful explanation, a subject's emotion-based appraisal of an artwork, treated *as if* it were an intrinsic property of the work itself. So *Mysteries of Udolpho* is not intrinsically boring, but rather I'm bored (by X) where the properties of X are incidental. In the history of aesthetics, this is what Theodor Lipps calls *Einfühlung* (translated as "empathy") or more literally "feeling-into."

4) *Prosthetic.* The object or scene extends our emotional range—for example, the sublime object makes us feel something we could not otherwise. Someone suffers in our stead as in the Passion of Christ (1 Peter 3:14, *King James* version henceforth): "But and if ye suffer for righteousness' sake, happy *are ye*: and be not afraid of their terror, neither be troubled; But sanctify the Lord God in your hearts: and *be* ready always to *give* an answer to every man that asketh you a reason of the hope that is in you with meekness and fear."[17]

5) *Performative.* As we have seen, William Reddy offers "I love you" as an *emotive* that does not merely describe some preexisting condition, it instantiates the emotion and it does so unpredictably because the outcome is not certain. So for Reddy, love literatures provide essential materials for a study on romantic love because that is where an emotive like the declaration of love appears through its material: its generic and social viability, its concrete opportunity, its deployment, its precise form, its receptions.

6) *Voluntary*. This, Philip Fisher argues in *The Vehement Passions*, is "where we feel something exactly because the other does not" (142).[18] A blank spot appears where the reader or spectator "*volunteers* passion, stepping in to supply the missing fear, grief, shame, or anger" (144). Fisher offers a typical Jane Austen move where we might feel shame for an action performed by someone like Darcy "shamelessly." Then finally and most importantly for my study:

7) *Systematic*. Any particular emotion or emotional experience—or for that matter, a marked *non*-emotional experience like shamelessness or apathy—appears only as it is part of a system. This can be a political economy of emotion like we get in Hobbes where emotions like fear and vainglory are distributed systematically but unevenly across the corporate body. Or it can be a "theory," as Ngai calls it (84), where feeling systematically produces the object of experience. For example, Sterne's *Sentimental Journey*, where the object of experience and a focalized theory of the sentimental are co-constitutive, such that the mechanisms of production are themselves apparent. Systematic considerations help us not only to account for emotions in their positive manifestation, but also to ask how emotions are superseded, excluded, or otherwise rendered unavailable or beyond the range of immediate experiential or discourse-analytic possibilities.

Thus talk about emotions can be confused for a number of reasons, and the nature-nurture debate is just one of the less interesting. First we get confused because the reference point can shift depending upon context and a set of related terms: feeling, mood, passion, affect, and emotion—the last two of which have been set against one another controversially as I will now explain briefly.

Considered historically, and occasionally in this book, "affect" can be simply a synonym for emotion. But from a transdisciplinary perspective, this terminology can be a point of inadvertent, but also sharp, disagreement about priority. Historically, "affect" can designate a general physiological disposition that precedes emotion theoretically, temporally, phylogenetically, and/or ontogenetically. In Shakespeare's *Love's Labour's Lost* I.i.149 we read "For euery man with his affectes is borne," but we would not expect to hear that we are born "angry" let alone "jealous," which are most definitely emotions in Shakespeare and beyond. From one polemic perspective with historical roots, affect comes first, and then we grow into emotions as individuals and as a species. This formulation in more recent scientific psychology considers affect the *experience* of a feeling or emotion, and hence it would be something (unlike emotion) that we share with lower organisms.[19] According to Ruth Leys, who has written polemically about this conceptual distinction, the definition of affect has recently consolidated across some neuroscience and also some high theory as "independent of, and in an important sense prior to, ideology—that is, prior to intentions, meanings, reasons, and beliefs—because they are nonsignifying, autonomic processes that take place below the threshold of conscious awareness and meaning."[20] Her analysis is based in part on a critical response to Brian Massumi,[21] which appears as the inadvertent return of Cartesian mind/body dualism where affect is associated with body, and emotion with the mind.

Practically, I approach these problems at the level of more local argumentation. What is the claim for one of these key terms, and which historical and argumentative warrants allow for such claims? What sorts of phenomena are at issue materially, and methodologically? Affect and emotion studies are difficult when one proceeds hamstrung, with one key term valorized and the other condemned; the best work avoids such a prior commitment. Here's an example with reference to an emotion already introduced: love. Lauren Berlant mobilizes the traditional emotion "love," as it plays an aspirational role for the Dardenne character Rosetta, while love's failure in that filmic context can be understood only according to the dynamic that Berlant calls post-Fordist "affect."[22] *Not* post-Fordist emotion. In this case "affect" better captures the lowness, as in "low affect," and also the negative space created by this particular pattern of failed aspiration which appears in the context of the 1980s Belgian working class and its Rosetta. That said, the space itself would never appear as such—could not be measured—without the traditional emotion "love" as a crucial but vanishing reference point in that context.

Emotion as a *reference point* is important, and it brings us back to the seminal analysis of Raymond Williams, who has had a significant impact on affect and emotion studies, including the work of Berlant. After Massumi and his influence, we would do well to revisit Williams if we want to work through the ideology of emotion historically, without simply condemning it as such. In *The Long Revolution*, for example, Williams presciently details how a "creditable emotion" comes to serve as a reference point and even as a lived experience that directly contradicts the conditions for its own satisfaction. And although this passage gets us away from the eighteenth century, it places an appropriate coda on a long historical arc that can be defined in terms of a new orientation toward well-being, including happiness and what Williams calls in this passage the "feeling of freedom":

> The very strong case for general planning, not simply to avoid waste but to promote essential development, research and reorganization, is practically nullified by a wholly creditable emotion: that we reject the idea of this kind of

> economic system controlling our lives. True, we are controlled now and will continue to be controlled by a quite different system, with its own denials and rigidities, but in the first place this is very much harder to identify, and secondly by its very structure and ideology, it appears to offer, and in just enough places does offer, the feeling of freedom.[23]

According to Williams, this crucial phenomenon of late liberal democracies—this feeling of freedom carefully controlled—would not be well served by the appeal to ideological superstructure alone, because it does not transpire at the level of ideas. Likewise, "hegemony" relies too heavily upon the sense of common sense to explain domination in the cultural domain. But neither is this creditable emotion pre-personal or pre-ideological in the sense of Massumi's affective turn, because it must be understood, substantially, in its *relationship to* the ideological, which is exactly where the word "freedom" refers. Finally, the feeling of freedom is not well served by systematic philosophy, à la Martha Nussbaum, because the "appearance to offer" and even the "just enough" of this creditable emotion is rhetorical. For it is precisely the structure and experience of the offer that matters in this case, not its systematic justification or even its normativity. Or to put this another way, a "creditable emotion" directly involved in political life—call it more traditionally a passion for freedom—refers to and actually embodies up to a point this cornerstone of liberal democracy, primarily as a carefully controlled alibi keeping at bay the welfare state in its socialist leanings.[24] Very much like Williams, the strength of Berlant's analysis depends upon the fact that she can work historically across these two interfering dynamics—affect and emotion—without dismissing emotion as a term somehow fallen from progressive political grace, and therefore tainted as a scholarly reference point. By way of contrast, theorists after Massumi who depend on affect as pre-ideological, presubjective, and prelinguistic, can wind up sounding mystical because they are not left with much to talk about.[25]

There is another reason we get unnecessarily confused in affect and emotion studies. Sometimes without realizing it, we bring to bear competing psychologies of emotion whether lay or expert. For example, the "James-Lange" theory, where physiological arousal instigates the experience of a specific emotion (crying makes me sad) conflicts with a cognitive appraisal model (poverty makes me sad).

Finally—and this is important—we get confused because "emotion" can be implicated in any one of the rhetorics listed above where there are probably others, and sometimes in more than one at the same time, depending upon the perspective rendered. Extending an earlier example, now consider

God's love. "Imitation of Christ" is *mimetic* as we are advised in Ephesians 5:2, "walk in love, as Christ also hath loved us." It is *prosthetic* as it makes us feel something we could not otherwise, and therefore John 4:19, "we love Him, because He first loved us." It is not descriptive, but rather *performative* when someone says "Jesus loves you," thereby instantiating the emotion unpredictably because the outcome is not certain. And finally it is *systematic*, as it appears, for instance, in Aquinas's systematic theology of the passions, or very differently in Equiano's spiritual autobiography, where we will read how the crucified Savior bleeding on the cross on Mount Calvary "unseals" the book and makes us read everything differently again, from the beginning (chapter 2). One goal of this book is to demonstrate how the rhetorical situation determines what we mean by "emotion" in the first place, and track its usage, which will be variable along these lines.

Methodologically, my rhetoric of emotion takes advantage of resources in that named humanities discipline. But to get a sense for the ways in which emotional discomfort has become a problem for us, we must widen the lens to include certain sorts of literary criticism and philosophy, which together point toward the European bourgeoisie and its legacy in sentimental fiction—but also in our own "everyday" scenarios like the initial dinner party fiasco, which is nothing if not bourgeois.

Discomfort with the Humanities

First my approach to literary criticism finds some good company as it asks how discomfort appears. Writing about social mistakes, Kent Puckett has shown how "bad form" works to produce coherence in social and textual systems, how a mistake like Charles Bovary's opening hat mishap makes us feel narrative form at work when embarrassment shapes character around bourgeois conventions violated, and shame provides the semi-anonymous voice that would implicate a reader's investment even if that investment ultimately takes the form of social criticism à la Flaubert: "Shame on you!"[26] In turn, what I call emotional "instability" hinges on a world that is social, but must be understood beyond the structures of social convention if the emotion is to appear in its proper place. Not Charles Bovary's embarrassment, but Yorick's "horror" is a defining sentiment we will revisit in later chapters, as it is tuned not to social convention, but rather to the environment where social life takes one form amongst others, including nonhuman. Bourgeois comfort is regularly challenged by sentimental narratives as they unfold—for example, Yorick's histrionics in Sterne's *Sentimental Journey*, including the way that he regularly botches "sentimental commerce" across rank, region, and gender.

At this point a role for literature, I offer, is to provide a detailed phenomenology of being-uncomfortable that adjusts situated theories of emotion and cognition as it has emerged in the philosophies of Dreyfus, Haugeland, Clark, Thompson, Wheeler, Noë, and others.[27] So sentimental literature provides in my study the substantial material for engagement with this important philosophical subfield, but a distinction is also drawn on properly philosophical grounds where I bring into focus the foundational metaphor.

If not the much-maligned "computational" metaphor—the brain is a computer and the mind is the program that the brain runs—what is the metaphor system of situated cognition theories now prominent in philosophy? We have already seen how, for Griffiths and Scarantino, the "cultural scaffolding of emotional performances" exceeds individual feelings and instead supports the right emotions at the right times necessary for society to "function *smoothly*" (italics are mine throughout this paragraph). And this metaphor system prevails in the philosophical literature. Setting the stage we have ethologists Jakob von Uexküll in the 1930s, launching his influential theory of the *Umwelt*: "we begin . . . with a subject located in its environment and research its *harmonious* relationships to the individual objects that present themselves."[28] More recently Andy Clark invokes the dance and jazz ensemble to illustrate how continuous reciprocal causation works across brain, body, and local environment (24); following Hubert Dreyfus, this fine arts dynamic is *fluid* (31); the environmental coping which might otherwise feel rough, *smooth* (as in smooth jazz, not Ornette Coleman or someone more challenging, I guess; see Wheeler, 194). Then exiting this languid stage and heading home for more domestic rest and relaxation, John Haugeland draws back the covers on a mind "*intimately embedded* in its world" (quoted in Clark, xxvi; Wheeler, 193) where, thoughtfully, Evan Thompson lays fresh linens, and I quote: "Culture is no mere external addition or support to cognition; it is *woven into the very fabric* of each human mind from the beginning" (403).[29] How did we get so comfortable, and says who? Read so, the literature on situated cognition appears on the edge of melancholia as it recuperates a world that would be much more comfortable—Freud might call this a canny return home—its rhetoric reassuring, as it allays our worst fears about being-out-of-place and feeling *unheimliche* as we are exiled from the familiar.[30] Dramatized and admittedly polemic, I imagine at this point a philosophical voice saying something like the following: "Thank God culture is woven into the very fabric of each human mind from the beginning because now I don't need to worry so much about those exaggerated problems of alienation." Referencing the title of his 2013 book and following Max Weber, literary critic Franco Moretti might call this a distinctly "bourgeois" form of philosophy that no doubt offers the threads of its own undoing.[31] But first let me follow the positive thread a bit further, because it helps us see some historical continuity in the modern period I'm treating.

Heading not toward philosophy but literature, Moretti quotes Weber from *The Protestant Ethic* as he discusses the new ascetic types of the northern European eighteenth century: "Over against the glitter and ostentation of feudal magnificence which, resting on an unsound economic basis, prefers a sordid elegance to a sober simplicity, they set the clean and solid comfort [*Bequemlichkeit*] of the middle-class home as an ideal."[32] In fact Moretti's third keyword *comfort* is embodied in the English bourgeois home and the historical sea change he identifies from the late seventeenth to the early eighteenth century when comfort is no longer what returns us to a normal state from adverse circumstances, but what "takes normality as its starting point *and pursues well-being as an end in itself*" (Moretti, 46, emphasis author's, offering the transitional 1719 example *Robinson Crusoe*). As opposed to bourgeois fashion that wants to resemble the old ruling class, comfort according to Moretti remains "down-to-earth, prosaic" (50), which indicates why we'll find it serving such purpose in sentimental literature. Moretti sums up the paradox of comfort on the way to explaining how a keyword can function: "during the seventeenth and eighteenth century, two equally powerful but completely contradictory sets of values came simultaneously into being: the ascetic imperative of modern production—and the desire for enjoyment of the rising social group. Comfort and *Genussmittel* [means of pleasure like coffee, tobacco, chocolate, spirits] managed to forge a compromise between these opposite forces" (51). But where Moretti sees comfort returning centuries later as relief from work, I read comfort returning to our intellectual scene where it provides a relief from the terrible stresses Sara Ahmed would call queer: social disorientation and alienation first among them.[33] Likewise, I will keep track of emotional labor as it takes shape unevenly; sentimental literature provides the scene of (emotional) labor concealed elsewhere including much philosophy, and hence it is a unique venue for analyzing a new kind of labor relations that constitute Enlightenment-era modernity. Or to put this another way, following Kenneth Burke, sentimental literature is where the terms if not the experience of bourgeois comfort are worked out and secured—literature as consolation.

In terms of philosophy, then, I want to unsettle situated cognition theory by reading its organizing metaphor against the alternative offered in some rather vexed literature—following Lukács in *Theory of the Novel*, Moretti writes every form is "the resolution of a fundamental dissonance of existence." But it turns out philosophy itself has something to offer in this regard. The queer version of critical theory finds some common cause in the 1930s Gestalt psychology of Kurt Lewin who wrote about the *hopeless* situation, and in Wittgenstein, for whom everyday aesthetic gestures take shape in the context of discontent, disgust, and "discomfort" (*Lectures on Aesthetics*, 1938). The uncomfortable situation, in other words, also has some deep philosophical roots exposed with a queer edge, so we can return to someone like Dreyfus on the leap of existence and find there an opportunity for new kinds of critical work.[34]

Or reversing the analysis, our comfortable philosopher takes shape in a form that speaks urgently to eighteenth-century European history and literature where comfort itself—now the feeling of well-being—became for the first time a way of life. I don't want to argue about how either comfort or discomfort is more basically human; at that level of generality both can be plausibly argued and hence the alternatives are uninteresting like any other general polemic around human nature. Instead the argument is historicist: the very distinction takes shape in a particular European eighteenth century where the economics of well-being were initiated for our age most obviously in utilitarianism (Austen, chapter 4), but also in key social distinctions where lived forms of well-being were shaped against degenerative and unaccountable forms that do not map seamlessly onto economics broadly understood. Hence the significance of literature where such forms of living (or not) were figured—not just as legal or philosophical or utilitarian abstractions, and not just as the idealized consequence of conduct books and education whether secular or religious—but amidst the more chaotic idioms of transaction. Cataloging the scenes of subjection around racial slavery I'll revisit in chapter 3, Saidiya V. Hartman contrasts a failure of post-abolition man with pedestrian practices like stealing away, the breakdown, moving about, pilfering, that occur below the threshold of formal equality. Similarly I am interested in how well-being and its bourgeois comforts—whatever the metrics of our recent happiness indexes might tell you—unravel in the literary artifact that can show you perfectly well the force of ideology, but also the proliferating forms of slippage and personal failure that tell you something crucial about how particular ways of being are formulated and where their limits reside.

Chapter Outlines (with a Definition of Rhetoric and a Defense of Fiction)

This book proceeds by way of a ground clearing gesture, with Darwin left standing, four case studies each exploring a different quality of the uncomfortable situation—loss of composure, bearing up, environmental hostility, feeling mixed—and an epilogue that offers, with ecological neuroscientist Stephanie Preston, a different kind of cross-disciplinary collaboration.

In chapter 1, I argue that Darwin's *The Expression of the Emotions in Man and Animals* can (also) serve as a foundational text for a humanities approach to emotion, but only if it is first disentangled from its heavy-handed editor Paul Ekman and his basic emotions program, which has done Darwin a profound disservice. So liberated, I then argue, Darwin's science of emotion provides a reference point for scholars in the humanities now trying to make their literary criticism speak to natural science. The photograph, illustration, and story-filled *Expression* (as Darwin called it) is both rhetorical and inseparable from its science that we sometimes imagine transcending its bookish material. In this chapter I recall how Darwin's *Expression* foregrounds the inherent rhetoricity of emotion, thereby outstripping Ekman's science of emotion that claims to follow in its wake, and which has recently found some advocates in the new subfield of Cognitive Approaches to Literature. Instead, I argue that Darwin's rhetoric of emotion is remarkably skeptical, which does not diminish its scientific piquancy, but rather aligns it with our situated theories in the science of cognition mobilized, among other places, by the philosopher of biology Alva Noë in his academic bestseller *Out of Our Heads: Why You Are Not Your Brain, and Other Lessons from the Biology of Consciousness* (2009). Thus realigned, I conclude, Darwin's *Expression* helps us understand anew how an adequate model of cognition must be able to account for literary emotion—a claim I detail in the following chapters. Schematically I contrast a more workable theory of situated emotion and cognition with 1) computational models of mind, and 2) models of mind and emotion in evolutionary psychology.

With the soil turned for a different story, chapter 1 also pursues the first of four case studies that show how a worthwhile science of emotion works in the study of literature, where sentimental literature offers a rich example. Obsessively interested in emotions as they take narrative form, sentimental literature around the turn of the European eighteenth century produced an unprecedented set of experiments that tested how "the man of feeling" along with a host of unstable characters might find composure—be composed—in a world where explanations for human activity tended instead toward new forms of social reason.[35]

Methodologically, chapter 2 offers a corrective to certain cognitive and Darwinist approaches to literature while developing a more detailed example of emotion, or affect criticism. "Bearing up in *The Interesting Narrative of the Life of Olaudah Equiano, or Gustavus Vassa, the African*" (1789) picks up a basic question: What on earth might it mean when Equiano writes he is "happily situated" as a slave? And how does this contradictory situation challenge our sciences of emotion? It would seem even in circumstances that are relatively hospitable, practically reconciled, and religiously committed to a scene that could sometimes appear God-given, slavery should under no description appear happy, or "comfortable" as Equiano later describes some of his enslaved fellows. Even the contrary *uncomfortable* is an understatement to the point where the very alternative marks the historic disjuncture noted by Moretti. Well-being is marked; it is not the universal that hedonic psychology sometimes studies at the level of the human organism tending toward homeostasis. Drawing resources from Equiano's slave narrative and critical race theory from W. E. B. Du Bois on double consciousness to recent Afro-pessimism (Moten, Sexton), this chapter explores what kind of narrative resources might be available to a science of dramatic displacement. By way of negativity, Equiano's *Interesting Narrative* shows in detail how well-being is systematically oriented (and incidentally why the "hostile work environment" mentioned in chapter 4 is defined according to legally protected classes like race, color, sex, family status, etc.). Where Saidiya Hartman worries about "the very effort to represent the situation of the subaltern" (10) because she is dependent upon what she calls the barbarous archive, I pursue in this chapter a related methodological concern for the "situation" under extreme philosophical pressure. In this racial slave situation, what counts simply as "being-there," and under what conditions? How do things appear and at what cost? How does Equiano feel in his own skin?

On first impression it would seem Equiano slips conveniently into our sciences of happiness—notably Daniel Kahneman et al.—and our theories of situated cognition drawing directly from Merleau-Ponty, as in the pivotal work of Alva Noë including *Out Of Our Heads*. Apparently Equiano adjusts to his situation. But is this really the story? My demonstration in this chapter proceeds through Equiano's time, his attachment/feeling, and his bearing up, while remaining attentive to the profound inconveniences that Equiano suffers despite whatever comfortable accommodations appear on the horizon. In each of these positive instances I'll show how Equiano's testimony speaks loudly to our sciences of happiness and situated cognition, before dissipating into queries and quibbles and counterarguments that should, in the end, threaten any easy sense of well-being. I will thus be working a local problematic at the intersection of science and the humanities, paying special attention to rhetorical qualities of a phenomenon—well-being—that escapes our scientific interests narrowly conceived. At the same time with Equiano's help, I'll attend to hedonic psychology and theories of situated cognition where resilient threads from the humanities might be reinforced. Both of these science-oriented projects, it turns out, have an impulse that is explicitly rhetorical, which means they wind up sharing significant ground with the humanities where rhetoric usually resides.

Now for the sake of definition, rhetorical theory has three dimensions in this book: 1) It alerts us to forms of address wherever they may appear; 2) it invites critical distance from those forms; and 3) it maps some of the important ways that the world already appears to us by way of rhetorical possibilities and prohibitions. In the spirit of Kenneth Burke, I explore in this book how rhetorical theory provides a historically rich field guide to the threatening and promising environments that are especially human, including the more narrowly defined "rhetorical situation" which has been considered by at least one key commentator fundamentally uncomfortable.[36] Rhetoric, as a humanistic form of affordance theory, posits a situation that is not neutral but is rather "persuasive," threatening and promising with respect to our human being-in-the-world.[37]

Here I am referencing Burke, the rhetorician and literary critic who, for instance, observes the following about eighteenth-century bourgeois poetry of the sublime, as it appeared from a perspective in the following century. I will quote at some length because the passage contains many of the elements in Burke that inspire my work in this project, thus indicating where I see my work fitting into rhetorical theory per se.

> The whole subject of "beauty" became obscured in much aesthetic theory of the nineteenth century because it tended to start from notions of *decoration* rather than from notions of the *sublime*. There are many possible ingredients behind this motivation, among them being the fact that aesthetic theorizing was largely done *by* people in comfortable situations for people in comfortable situations. But there is a subtler factor operating here: poetry is produced for purposes of comfort, as part of the *consolatio philosophiae*. It is undertaken as equipment for living, as a ritualistic way of arming us to confront perplexities and risks. It would *protect* us. Let us remind ourselves, however, that implicit in the idea of protection there is the idea of something to be *protected against*. Hence, to analyze the element of *comfort* in beauty, without false emphasis, we must be less monistic, more "dialectical," in that we include also, as an important aspect of the recipe, the element of *discomfort* (actual or threatened) for which the poetry is "medicine," therapeutic or prophylactic.[38]

As I read this passage rather selfishly, Burke is making a point about how we must approach the rhetoric of emotion historically. We must always ask who is attributing which emotions to whom—bracket sincerity—and what this act of attribution is supposed to accomplish. So for example in eighteenth-century sentimental literature including some works of what Burke calls the poetry of the sublime, authorial attribution of emotion to the "man of feeling"—like Yorick for instance—says little about his sincere emotions whatever that might mean, and quite a lot about the "discomfort" that would be addressed by this very attribution. Similarly Equiano's self-attribution of emotion and then our reading of it, which itself has distinct historical contours, must be considered rhetorically as a form of address and not just as sincere expression. So in some basic ways I draw from Burke's literature as equipment for living, and then Burke's rhetoric as a field guide to the sort of lives that might be lived.

Dimension one of my rhetorical theory therefore points to effectiveness, which is to say Equiano's happy situation appears in the narrative consequentially; it does not describe first and foremost. Its design to do something is controlled only partially by authorial intention and is discernible by the reader only at the risk of committing the intentional fallacy. Instead most obviously, the passage about his happy situation asks for some kind of a response from the reader, but a response irreducible to simple identification understood as suffering-with. For instance, a reader moved to abolition by Equiano's narrative need not feel happy or miserable with Equiano, whom we are not sure feels one way or another anyway: he qualifies, ironizes, looks back and forth with art leading the way. Let alone the reader, or the inverted reader: Aristotle understood full well that tragedy can be a source of pleasure, for instance. The design itself carries rhetorical force by inhabiting conventions. At the same time, the author Equiano figures the convention ironically, which only makes sense in relation to his life story, which is to say you have to read the story or at least know the setup for the irony to appear. In other words, the emotion is "rhetorical" in that it requires, to appear as such, readers and audiences, which is where effectiveness finds its orientation.

Second, "rhetoric" in this case means the emotion—*pathos* in the classical lexicon—is not first and foremost something Equiano feels, such as happy, nor is it something observed like a smile or a positive disposition or an authorial attitude, but instead it transacts. Equiano does not feel happy but instead finds himself, ironically, in a happy situation that posits an attitude toward the scene of subjection (~terror), the master-slave relationship (~gratitude), and a type of social identity (~quite English). In other words, this happy situation would be lost on anyone who isn't paying attention—again it is rhetorical—and it cannot function primarily as an assessable state-description. A rhetorical understanding of emotion thus offers not only a critique of scientific theories that can't adequately take the situation into account; it also nuances the "history of emotion" à la Reddy and Rosenwein as it focuses on social transactions.

Finally, rhetoric as a field guide: traditionally next to deliberative rhetoric where the goal is audience activation, we find forensic rhetoric where the goal is ultimately judgment, and epideictic rhetoric (ceremonial praise or blame) where the goal is affecting attitudes. In fact rhetorical theory traditionally goes into great detail about how the general vectors of threat and promise can be structured for maximum effect: arrange thus and so, deliver not like this but like that. Thus rhetoric, as a humanistic form of affordance theory—see Gibson below—posits a situation that is not neutral but is rather "persuasive." Or to put this another way, rhetorical theory enhances awareness specially appropriate to human environments.

Chapter 3 continues to make the case for a literary mode—the sentimental—while explaining what sort of things happen when the genre shifts from literary nonfiction—that is, the slave narrative—to the Gothic novel. "Hostile Environments in Ann Radcliffe's *The Romance of the Forest*" (1791) provides the next case study where this time the environments, technically understood with the help of perceptual psychologist J. J. Gibson, make people terribly uncomfortable to the point of becoming unhinged. Or even "undone," as rape victim Antonia protests in Matthew Lewis's *The Monk* (1796) just before narrative logic finishes her off: "Is not my ruin completed? Am I not undone, undone for ever?" In this chapter I show how the cause of death in this instance, and more generally the terms of a life undone, find their genre in the sentimental version of Gothic fiction where the emotional scene is vexed. The sensations of princes and heroes "do not stamp the same affections on their domestics," observes an early practitioner Horace Walpole, and from this observation about the vexed emotional scene derives a principle for narrative construction Walpole finds first in Voltaire's *mélange*: "Nothing is more common than a house in which a father is scolding a daughter—absorbed in her emotions—weeping; the son makes fun of both of them, some relatives take different sides in this scene, etc." (quoted in *The Castle of Otranto* preface to the second edition, 1765). A narrative logic Walpole, Lewis, and especially Radcliffe pursue with notable success shows how—in contrast to tragedy, which undoes through character—such domestic "scenes" and other human environments can pose a real threat.

Not surprisingly this chapter on the environment serves as a particular kind of ecocriticism, as we now call it, while at the same time energetically distinguishing itself from "literary Darwinism" and related versions of literary criticism where the ancient environment of evolutionary psychology exerts undue pressure on the scene. Here instead of evolutionary psychology, the perceptual psychology of Gibson provides the basic definition of an environment. I then adjust Gibson's definition toward a humanities perspective where Gibson's invariance criterion hinders our understanding of a hostile environment first in legal, and then in literary terms where, I argue, interpretive patterns stake their own claims. Antonia's hostile environment, for instance, is *not* invariant with respect to human beings in general and we should ask how so, mapping carefully the patterns and consequences that define the hostile environment in this case. I also take time in this chapter to show how some literary criticism that falls under the heading of Cognitive Approaches—including work by Elaine Scarry, cognitive rhetoric à la George Lakoff, and David Herman who practices *Umwelt* research—already benefits from an environmental approach referencing Gibson.

Thus the state of the art: it matters very much what model of cognition one brings to bear on the study of literature, in the laboratory, and beyond. First and foremost my effort can be understood as a response to students and colleagues 1) dissatisfied with cognitive approaches to literature as they sometimes rely upon a version of the computational model of mind, and 2) intrigued by situated cognition theory that—other than a flicker here and there including a nice essay by David Herman—has not much been applied to the study of literature, or to the other humanities and arts for that matter. This despite the fact that situated cognition theory holds out relatively more promise for work in the humanities, because it must pay very close attention to the proximate (not Paleolithic) environments where cognition appears and which means, in turn, that the precise forms of that appearance—historical and aesthetic—are central. Folks in the humanities care very much about the whats and wherefores of aesthetic form. And thus I offer one simple way to describe this book: it is my effort to fill a methodological gap by applying situated cognition and emotion theory to literature. That said, I first found myself attracted to this project because it could be done best by way of my home field, rhetoric—the traditional humanities field where "situations" have been explicitly theorized the longest, and in depth.

So chapter 3 on *The Romance of the Forest* expands our understanding of how emotion works in the rhetorical situation. The larger point is not that emotions in the laboratory suffer a thin description while emotions in the wild enjoy a thick description especially when they appear in narrative literature, which offers an emotional situation in detail. Ultimately after critique, my methodological point is comparative. Laboratory work on emotion affords a certain kind of emotional situation where variables must be identified and for the most part, held constant. The experiments frame the emotions, which appear in the subjects primed in a certain way and represented in the data as interpreted; the situation is best set up to capture what I call in my typology *mimetic* emotion. Literary nonfiction such as the slave narrative affords another kind of emotional situation where variables like "happiness" might be identified but without the same kind of constancy requirements; most importantly they stake a claim or make a point or appeal to the reader, which is to say they are manifestly rhetorical. They appear only incidentally in the subject, and most evidently in a textual apparatus understood historically—the effort to hold constant does not apply in the same way, and interpretive variance over audiences and over time need not be a disadvantage.

Likewise literary fiction—for example, a Gothic novel in the sentimental mode—offers a rhetoric of emotion although the frame affords something importantly different, namely attention to the frame *as such* (remember for instance Sterne's "commentary on the text"). In its manifest detachment from actual situations, literary fiction allows us to better focus on emotional form by drawing attention to its rhetoric without deferring to practical exigency—in a way the novel serves as an emotional laboratory of a different sort where possibilities can be explored without giving way to probabilities. The hostile environment that appears in Radcliffe's *The Romance of the Forest* depends upon many things but not some kind of practical threat that would in fact diminish the formal project and thus the emotion in its more precise determination. Fictional status allows for emotion studies not available otherwise. Adeline is the fictional heroine threatened virtually by the stuff of imagination, like a hollow sigh that seems to pass near her at one point—but the light of day dispels her irrational fear. Subsequently she is actually threatened by the Marquis, who is also, of course, the stuff of imagination on many levels, including on the level of "fiction" that provides the novel its basic conceit. The Marquis is a fiction. But is he? In a formula: real terror, the novel teaches us in this instance, is fictional. Terror, unlike a dangerous situation, is the stuff of uncertainties as it takes provisional shape and is characterized with respect to our fragile sense of well-being. Hence also the centrality of rhetoric per se, as human being is done up in certain ways—terror is irreducibly a form of address.

Or to put this in terms of philosophy of language: the threat to Adeline is not real, nor is Antonia's rape, which is to say these people never existed—they are nonreferring terms in the world of fiction. But does that mean the abstracted sentence "Antonia was raped" or even "X was raped" is unaffecting? Not necessarily. Like the dinner party scenario, in this case what matters is not the reference alone—though of course that reference can make quite a difference practically speaking and hence the considerations of literary nonfiction or simple reporting by police for instance—but rather the rhetoric of possibility that embeds a sentence like this in the world beyond any immediate practicalities. The very fact that "X was raped" serves as a grammatically correct, semantically meaningful, and practically possible sentence (of a horrific sort) makes it affecting. What the world does with a sentence like this ranges quite broadly, however, and likewise what an author of sentimental fiction does with this sentence form or its neighbors like Antonia "being undone" can range widely as well. This is why the sentence form in Lewis or nearly in Radcliffe can be read meaningfully down to the level of the specific text or even the scene: Adeline was not raped after all but her very real fears are exquisitely mapped by Radcliffe against very real threats. Adeline's status as a fictional character thus has the added advantage over nonfiction in that she brings to bear, rigorously, the affectability of nonreferring terms.[39]

Now to summarize the case for fiction across three examples. Barsalou's laboratory anger, Equiano's nonfictional happiness, Adeline's imagined terror: Emotion Studies give us unique access to the very question of fictionality that cuts across all three. Or to put this another way, anyone interested in Emotion Studies—including a scientist—can learn something essential from literary fiction, for it is the very best domain for studying the emotional situation essential to the experience, along with the concrete forms that bring it to bear. There is good reason why subjects in the Barsalou laboratory familiarize themselves with mini-narratives that sound like they fell out of Craft of Fiction 101.

Finally these forms will vary, which is why research in Emotion Studies must transpire in a variety of literatures subsidiary to a generic category like the novel, or a modal category like the sentimental; for example, emotional form in Austen's *Sense and Sensibility* is "mixed" along lines that certainly speak to the traditional literary mode referenced in the title, while forming situations that can be exemplary for Emotion Studies writ large, including, as we will see, a scientific perspective. At the same time, this claim about varieties of emotional situations doesn't posit radical incommensurability (see epilogue). Since the phenomenon at hand, such as happiness, or terror, is manifestly shared across academic disciplines and domains of everyday experience, we are obligated to work carefully across some traditional boundaries. Meanwhile recent scientific work on irrationalism find a subtle advocate—perhaps even a sophisticated interlocutor—in the form of literature.

Mixed feelings, the topic of Chapter 4, are famously at the heart of Jane Austen's novel *Sense and Sensibility* (1811) and we think we know what that means, like a recipe: combine two parts a feeling like gratitude, one part happiness, a dash of resentment and you get something like Elinor. But mixed feelings in the novel and beyond are poorly served by this disequilibrium model recalling the long tradition back to Galen; in fact mixed feelings are a matter of negotiated circumstances where feelings may be at odds as they converge on character.[40] Hence the significance of literature and particularly the sentimental novel as a cross-disciplinary research domain, where this kind of rhetorical situation takes shape along lines that can be anchored beyond the novel per se. Referencing the science of situated emotion and hedonic psychology (Kahneman on bias and persuasion), this chapter offers a new way to understand mixed feelings as rhetorically situated. The novel opens with the Dashwood women hopelessly displaced from their Sussex home because in this case and in many others like it, a woman's well-being depends upon the diminishing returns of a man's benevolence: thus Mary Wollstonecraft's "equivocal beings" where sisters and then sisters-in-law are structurally at odds around the sentiments of a man. Says Fanny to Mr. Dashwood: "Do but consider, my dear Mr. Dashwood, how excessively comfortable your mother-in-law and her daughters may live on the interest of seven thousand pounds." By what means might I, the reader, establish what's going on around this "excessive comfort" that posits a norm framed by oxymoronic humor? Though fiction, this is once again a question of very real-world rhetoric, as I'll take time to explain further by way of Kahneman and Miller's 1986 work on "Norm Theory: Comparing Reality to Its Alternatives."[41] This work in psychological science shows how emotions appear in the rhetorical effort to establish norms, like the comfortable living owed Mr. Dashwood's women. What counts as comfortable in this case? Says who? Not only will the norm depend upon how different positions are brought to bear, in this case by the characters Fanny, Mr. Dashwood, and then also the Dashwood women, who are absent from the scene diegetically but very present by way of the narrative position that casts doubt upon Fanny's goodwill. The humor in this case depends upon a distribution of sympathies that casts the feeling of the moment—care or charity—outside the frame of the characters who express it, while pointing with someone like Wollstonecraft to a real-world situation that belies the immediate sentiment. As Kahneman and Miller demonstrate in ways that are surprisingly relevant to our literary example, "reality" is established with respect to its alternatives. In particular Kahneman and Miller study how sympathy is a function of the perceived abnormality of a victim's fate (What do these women actually deserve? What does anyone deserve?), the availability of counterfactuals (the good life appears just out of reach: not too close and not too distant), and the actions of a focal individual that seem mutable in comparison to background actors (Austen's art depends in part upon the expectation that Elinor bear some responsibility–focal characters map the activity domain disproportionately). Then on top of this baseline sympathy that is established in a network of nonreasonable expectations, Kahneman and Miller posit a correlation between the perceived abnormality of an event and "the intensity of the affective reaction to it, whether the affective reaction be one of regret, horror, or outrage" (146). Or in our case, the regret, horror, or outrage a reader might find in *Sense and Sensibility* depends upon norms like resource management that carry historical weight, as they are perceived sometimes shockingly through the events as narrated.

Thus chapter 4 also explores a methodological opportunity at the interface of science and the humanities, beyond some recent work in Cognitive Approaches that proceeds unecologically as it pursues—frequently with Austen—"theory of mind," or what is colloquially called mind reading, where the extra-mental falls away as mere projection or anthropomorphism. No doubt mind reading is one of the Austen dynamics, as critic Lisa Zunshine has readily shown. But in order to understand what's going on emotionally, the operative world, including something like resource distribution, must factor in, much like Kahneman would remind us, not as mental projection but rather as the real-world anchors that are established rhetorically as norms are negotiated with horrific and sometimes humorous effects.

Finally the epilogue, "Irreconcilable Differences?" explores what collaborative work across emotion science and the humanities might look like, as I compose with University of Michigan ecological neuroscientist Stephanie Preston.[42] By first outlining various limitations and constraints of laboratory science, we highlight some recent hardening of the "two cultures" divide in emotion research. Then we discuss experiments that show how laboratory research on emotions can be modified by the critical and rhetorical methods that distinguish the humanities. Classic laboratory studies of fear, for example, document the persistence of unconscious white vs. black bias while, at the same time, neglecting circumstances like the history of racial slavery that could provide causal explanations of just the sort that these same scientists first raise, then anxiously defer. More promising from a humanities perspective are experiments that draw on a critically informed, situated emotions approach, which explicitly incorporates language and audience considerations into experimental design, and into interpretation.

The hope is that more work on emotion is possible at the intersection of science and the humanities. But first we need some methodological adjustments. The study before you shows what those adjustments might look like.

1

Defending the Humanities with Charles Darwin's *The Expression of the Emotions in Man and Animals*

When Charles Darwin first published *The Expression of the Emotions in Man and Animals* (1872) his purpose was to show continuity in emotional expression across species and thereby strengthen his theory of evolution. Surrounding the book's illustrations, therefore, one finds descriptions of animal emotion designed to elicit our recognition as members of a universal audience. In this case (Fig. 1.1) our charm instantiates Darwin's argument for emotional continuity across species at the same time that we as members of an historical audience may miss the sly provocation of contemporaries with which the passage concludes.

> The lips of young orangs and chimpanzees are protruded, sometimes to a wonderful degree, under various circumstances. They act thus, not only when slightly angered, sulky, or disappointed, but when alarmed at anything—in one instance, at the sight of a turtle—and likewise when pleased. . . . The accompanying drawing represents a chimpanzee made sulky by an orange having been offered him, and then taken away. A similar protrusion or pouting of the lips, though to a much slighter degree, may be seen in sulky children.[1]

This passage, like many others in the *Expression* (as Darwin called it) is both rhetorical and inseparable from its science that we sometimes imagine transcending the material that binds it to accidents of medium, occasion, and person. In this chapter, I recall how Darwin's *Expression* foregrounds the inherent rhetoricity of emotion, thereby outstripping Paul Ekman's science of emotion that claims to follow in its wake, and which has recently infiltrated the humanities like a Trojan horse settling in a new critical subfield of Cognitive Approaches to Literature (CAL). Instead I will argue Darwin's rhetoric of emotion is remarkably skeptical and humanistic, which does not diminish its

FIGURE 1.1. "Chimpanzee disappointed and sulky. Drawn from life by Mr. Wood." From Charles Darwin, *The Expression of the Emotions in Man and Animals* (London: J. Murray, 1872), 139. (Wellcome Library, London, CC BY 4.0)

scientific piquancy, but rather aligns it with our "situated" theories in the science of cognition recently mobilized, among other places, by the philosopher of biology Alva Noë in *Out of Our Heads: Why You Are Not Your Brain, and Other Lessons from the Biology of Consciousness.*[2]

Though initially a bestseller, the *Expression* lapsed into relative obscurity during the next century as Darwin's evolutionary theory established itself primarily on other terms, including, most importantly, the fossil record, homologies across related life-forms, geographic distribution of related species, and artificial selection like dog breeding. Meanwhile, the ambiguities of studying emotion rendered it a difficult and even suspect science for the next century, especially insofar as the mechanics of our emotional life lacked the reliable metric promised by a rational life that might be reconstructed in the spirit of logical positivism or tracked, for instance, through idealized behaviors such as rational choice in the marketplace.[3] "It's raining" seemed the kind of thought that might reliably link the mind and the real world, but something like the experience of fear suffered from all sorts of semantic and practical ambiguities that made any experimental project difficult to realize, even after Darwin's evolutionary argument. More recently, however, the *Expression* has made a dramatic comeback as the study of emotion mushrooms across academic disciplines and in applied fields such as homeland security[4] or in a popular arena such as the cop drama *Lie to Me*, which was built around expert analysis of emotional microexpressions à la Ekman.[5] Now considered wrong about certain facts, such as the Lamarckian inheritance of acquired characteristics and the emotional primitivism of the insane,[6] Darwin's *Expression* has recently reemerged as the foundational work in the science of emotion, both in terms of methodology and theory.[7] In 1998, nearly a full century after the second edition, Ekman introduced and provided a running commentary on the third edition, underscoring how passages such as the following grounded the methodology for studying emotion still followed today across the social and natural sciences.

> Dr. Duchenne galvanized, as we have already seen [Fig. 1.2, 1.4], certain muscles in the face of an old man, whose skin was little sensitive, and thus produced various expressions which were photographed on a large scale. It fortunately occurred to me to show several of the best plates [Fig. 1.3, bottom], without a word of explanation, to above twenty educated persons of various ages and both sexes, asking them, in each case, by what emotion or feeling the old man was supposed to be agitated; and I recorded their answers in the words which they used. Several of the expressions were instantly recognized by almost every one, though described in not exactly the same terms; and these may, I think, be relied on as truthful, and will hereafter be specified. (*E*, p. 21)[8]

Then Ekman abruptly inserts the following comment in the 1998 edition: "Darwin describes here his use of a method to find out what information is conveyed by an expression: asking people to judge the emotion shown in a photograph without any information about the situation in which the expression occurred, and determining whether or not they agree. This has become the most commonly used method for studying facial expression, although Darwin is rarely cited as the first to use it."[9] Indeed, Ekman himself has applied what he considers essentially the same methodology over the last thirty-five years as he developed his influential work on the basic pancultural emotions displayed in the image set (Fig. 1.5). Can you name them?

Relying on this image set and others like it, Ekman has produced a methodology for studying emotion scientifically, and from this work he has developed the Facial Action Coding System (FACS) used widely in cognitive science designed to locate emotional centers in the brain by correlating brain

FIGURE 1.2. "Horror and Agony. Copied from a photograph by Dr. Duchenne." From Charles Darwin, *The Expression of the Emotions in Man and Animals* (London: J. Murray, 1872), 307. Duchenne's original dramatization gives us "une expression d'effroi mêlé de douleur extreme . . . à l'expression de cette terrible émotion de l'âme s'ajoute celle de la douleur horrible de son supplice. Cette expression doit être celle du damné" (Guillaume-Benjamin-Amand Duchenne de Boulogne, *Mécanisme de la physionomie humaine, ou analyse électro-physiologique de ses différents modes de l'expression* [Paris, 1862], p. 108). R. Andrew Cuthbertson translates this passage as "an expression of terror mixed with extreme pain . . . the harmful pain of torture has been added to the expression of this terrible emotion. This expression must be that of the damned" (Duchenne, *The Mechanism of Human Facial Expression: Studies in Emotion and Social Interaction*, trans. R. Andrew Cuthbertson [Cambridge: Cambridge University Press, 1990], 92). (Wellcome Library, London, CC BY 4.0)

activity with exposure to carefully selected photographs.[10] Of the cognitive scientists who rely upon FACS, Antonio Damasio is the most influential in the humanities, as we will see below.

In *Nature*, for example, Damasio describes an experiment where patients with a damaged amygdala judge faces on a trustworthiness and approachability scale.[11] The faces had been previously selected by "normal individuals" who were asked: "How would you rate this face on the scale of one to five, relative to the trustworthiness and approachability that the owner of the face

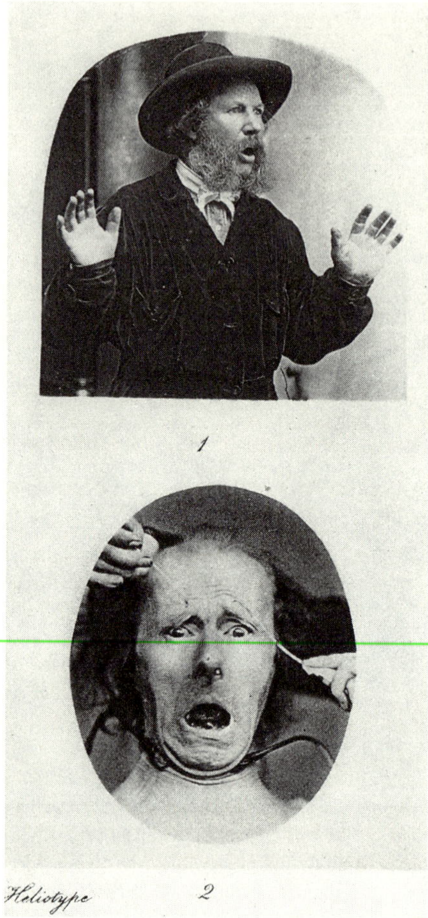

FIGURE 1.3. "The original photograph was shown to twenty-four persons, and they were separately asked, without any explanation being given, what expression was intended: twenty instantly answered, 'intense fright' or 'horror'; three said pain, and one extreme discomfort." From Charles Darwin, *The Expression of the Emotions in Man and Animals* (London: J. Murray, 1872), 299. The top image, which also appeared in the first edition of *Expression*, is of the famous London photographer Oscar Gustav Rejlander representing the emotion of "astonishment." (Wellcome Library, London, CC BY 4.0)

inspires? Or, in other words, how eager would you be to approach the person with this particular face if you needed help?" As it turns out, Damasio reports that amygdala-damaged patients judged as trustworthy and approachable faces "you or I" would consider suspicious and try to avoid.[12] Then on the basis of this experiment, Damasio and his colleagues located the relevant emotion in the brain, while bracketing some pressing questions about what counts as a trustworthy and approachable face, to whom, under what circumstances, and why. Technical issues abound.

In a section of his book provocatively titled "The New Phrenology?," Noë outlines a debilitating set of problems faced by researchers who try to locate us in our brains, including most importantly the following: 1) the impossibility of eliminating feedback produced by the two-directionality of the brain-senses loop; 2) normalizing imperatives that produce results against a stock brain; and 3) the impossibility of eliciting direct information about

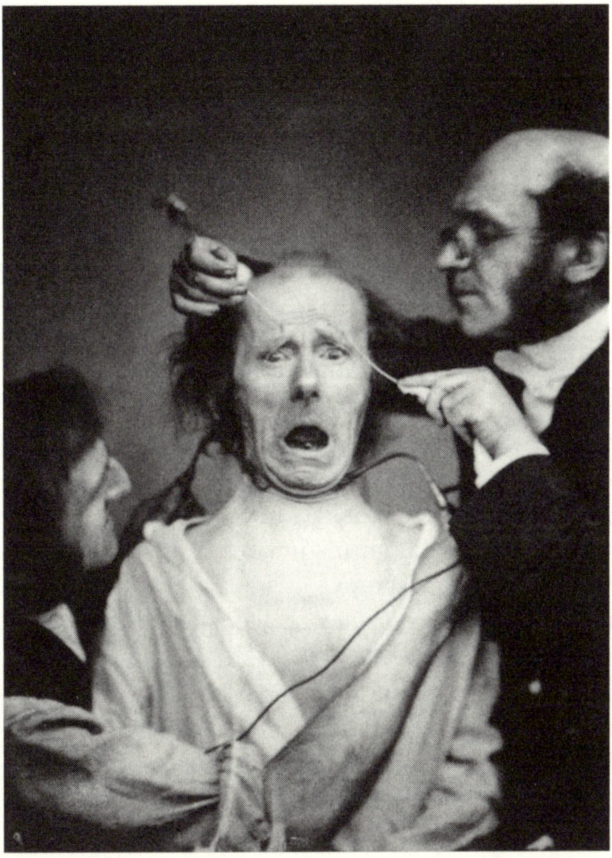

FIGURE 1.4. The uncropped Duchenne photograph. (Wellcome Library, London, CC BY 4.0)

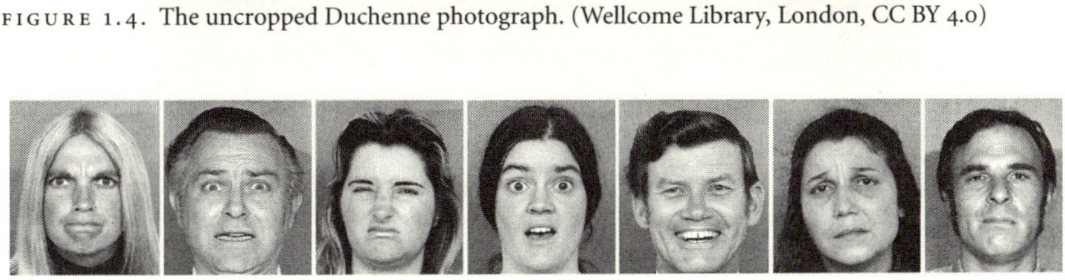

FIGURE 1.5. The facial expressions of Ekman's initial, basic emotions, from left to right, anger, fear, disgust, surprise, joy, sadness, and Ekman himself as contempt. (Reproduced with permission from Paul Ekman Group, LLC)

consciousness or cognition through PET and fMRI technologies that have to correlate physical magnitudes to blood flow, blood flow to neural activity, and finally neural activity to mental activity like feeling suspicious (see Noë, 19–24; also my epilogue). Although a brain-scan image may contain important information about neural activity related to a cognitive process, cautions Noë, "we need to take care not to be misled by the visual, pictorial character of these images. Brain scans are not pictures of cognitive processes in the brain in action" (Noë, 24).[13]

But for our purposes the shortcoming that matters most is methodological: emotions cannot be broken down into basic units—whether that means molecules,[14] brain images, or facial expressions—without losing track of the phenomenon at hand. Psychologist Jerome Kagan emphasizes our category mistake when the amygdala is activated by exposure to a photograph of an angry face and the resulting image is called an emotion. The neuroscientist who insists that a particular brain state represents an emotion resembles the physicist who denies the reality of my pen because the mathematical interpretation of quantum mechanics rejects the existence of stable objects.[15] Though it may be useful to look for the elementary foundations of both psychological and physical phenomena, Kagan concludes, complex phenomena possess distinct properties and therefore require "distinct metrics and principles" (*What Is Emotion?* 214). That Kagan's valedictorian address to his students and colleagues requires this reminder is not surprising.[16] Surprising is that cautionary tales about parsimony in human affairs speak to a growing number of humanities scholars who overlook precisely the situational nuances that would seem most amenable to their indigenous methodologies, as we will see below.

So what does an evolutionary psychologist like Ekman mean by basic emotions? Ekman applies the word *basic* to emotions evolved for their adaptive value in dealing with fundamental life tasks such as achievements, losses, frustrations, and so on. According to Ekman, each emotion "prompts us in a direction which, in the course of evolution, has done better than other solutions in recurring circumstances that are relevant to goals," and "innate factors play a role in accounting for the characteristics they share, not species-constant or species-variable learning."[17] Thus his commentary on Darwin's *Expression* winds down with the following strong program: "Although the specific event varies—the type of food, the general theme—ingesting something repulsive as a cause for disgust, or ingesting something attractive as a cause of enjoyment—is universal. *I think this is a good model for all the emotions.*"[18] However, this is also where Ekman and his legions leave Darwin behind, with a sleight of hand that obfuscates the subtlety of Darwin's work at the same time that it nudges to the periphery of serious science the emotions that escape this expressivist model of evolutionary biophysiology. In fact, as we will see below, Ekman doesn't consider anything beyond the basic emotions to be emotions at all.

Among other things, this impoverished Darwinian model suggests we investigate how a novel might serve as an emotionally competent object where certain formal features elicit from the reader a sympathetic or antipathetic response divorced from immediate action.[19]

An early example of such criticism would be June Howard's *American Literary History* article "What is Sentimentality?," which mobilizes Ekman and Damasio in an effort to undermine debates about whether sentimentality is a good or bad thing. More important than this evaluation, argues Howard, is a transdisciplinary investigation of sentimentality that would better explain how reading sentimental fiction is a bodily act where the text that produces "pulse beats and sobs . . . radically contracts the distance between narrated events and the moment of their reading, as the feelings in the story are made tangibly present in the flesh of the reader."[20] Likewise in *Shakespeare's Brain: Reading with Cognitive Theory*, Mary Thomas Crane draws from Ekman and Damasio to argue for a reading of literature that grounds the mind—including conscious and unconscious mental experiences of perception, thought, and language—in the brain and other bodily systems.[21] In Jane F. Thrailkill's *Affecting Fictions: Mind, Body, and Emotion in American Literary Realism*, Ekman and Damasio are mobilized throughout, particularly where the novel is described as an aesthetic technology of the human organism.[22] "In the relatively recent field known as Cognitive Approaches to Literary Studies," summarizes Suzanne Keen in her study of empathy and the novel, the work of Ekman, Damasio, and Joseph LeDoux have "virtually canonical status" insofar as matters of affect are generally considered under the umbrella term "*cognitive.*"[23] And that's where the impoverished Darwinian model promoted by Ekman and Damasio obscures more than it reveals (I critique "literary Darwinism" in chapter 4). A robust account of the social world is essential for our understanding of emotion, and therefore we are led astray by any parsimonious account of the social world where cognition is reduced to a bodily function. Instead, with Darwin I want to foreground a more appropriate rhetorical model of cognition where consciousness is situated in the brain-body-world nexus.

That's not to say cognitive approaches are impossible—indeed I offer my own alternative in this book which has certain affinities with the recent work of Terence Cave[24] and Jonathan Kramnick,[25] as well as Elaine Scarry and David Herman as I noted in the introduction. The issue is which cognitive approach, and which model of cognition. The critical works just mentioned and others like it might provide sensitive readings despite the impoverished model of consciousness at the core, but the returns are diminishing. In her lexical analysis of Shakespeare's plays (for instance, *house* and *home* in *The Comedy of Errors*), Mary Thomas Crane mobilizes the cognitive rhetorics of George Lakoff and Mark Turner while offering "a possible background" for Patricia Parker's cultural criticism of Shakespearean figures (*SB*, 33).[26] But precisely at the point where Crane situates such familiar linguistic analyses against a cognitive science background the argument fails. Though a critic like Gail Kern Paster might productively scrutinize bodily experience in relation to discourse, she has, according to Crane, paid relatively little attention to the brain, "the material place within the body where discourse is processed and therefore where discursive construction, if it occurs, must be located" (*SB*, 7). But everything humans do can be located in the brain at some point, so the observation does not help us read Shakespeare any more than it helps us drive to the grocery store, which we also do thanks, in part, to our brain. Crane explains far too much with the brain: it is the "material site where discourse enters the body, where entry into the symbolic occurs, and therefore where the subject is constructed" (*SB*, 7). It is "is a material basis for a limited sense of 'essential' human attributes as well a space for individual arrangements of neurons" (*SB*, 23). So when it comes time to discuss Timothy Bright and Shakespeare on the processes behind emotional expression, for instance, Ekman lurks predictably in the background (see *SB*, 244n19), foreclosing cultural criticism in the spirit of Parker, except as a supplement. And what does literature do for us? The brain, according to Crane, "constitutes the material site where biology *engages* culture to produce the mind and its manifestation, the text" (*SB*, 35; emphasis added). Mind is thereby sequestered in a brain that recedes from culture to the point where a chasm must be bridged; hence the palliative work of literature and its criticism.

This brain-world dichotomy is common in cognitive approaches to literature, which provide the ostensible cure. After insightful analysis of the circus scene from *The Adventures of Huckleberry Finn*, where Twain invites the reader to reflect unsentimentally on his or her emotional experience as it unfolds (the work of "realism"), Thrailkill concludes with a metaphor of integrity assured by the surgical work of literature and its criticism: "Works of literary realism help us to realize that, when we read, we are all like Narcissus staring into a puddle: delighting in that 'extra you' who, far from being a solipsistic illusion, is a neurologically nested affective companion keeping us from our isolation by suturing us, body and mind, firmly to ourselves and to the world in which we live."[27] And not surprisingly for Thrailkill this affective-cognitive achievement of literature relies explicitly on biological stability according to Darwin's *Expression* as filtered through Ekman: "Drawing in evolutionary theory, these emerging accounts suggest that while the cultural significance of feelings, along with rules about displaying them, may fluctuate over time, the actual corporeal architecture of emotional experience—almost universal to members of a species—has evolved so slowly over the course of millennia as to be, in a limited time frame of human history, practically stable."[28] This, in turn, means the literary critic like Thrailkill can acknowledge cultural criticism as important but secondary, and then trump it with biological criticism with ostensibly wider reach. But Darwin himself is an equivocal ally in this project.

In this case Darwin's genius lies partially in his comparative study of emotion across culture and species that avoids the dramatic reductions Ekman and other scientists of emotion rely upon to make their work acceptable within protocol defined too narrowly—reductions that later justify scientistic conclusions drawn too broadly. Though these humanistic commitments to scientific responsibility typically recognize some contributions of critical theory, they tip the balance by taking seriously advances in evolutionary psychology and cognitive science of the particular sort I have briefly outlined. But in reducing the domain of cultural explanation they achieve, whether intentionally or not, a dubious political end of the sort critiqued by Lauren Berlant in the introduction to her collection of essays *Compassion: The Culture and Politics of an Emotion.* When emotion is reduced to a basic unit, such as the face, we should always ask about the institutional consequences; in Berlant's case she tracks the trade-off when compassionate conservatism relocates a particular zone of intimacy.

> Compassionate conservatism advocates a sense of dignity to be derived from labor itself—of a particular sort. No longer casting a living wage, public education, affordable housing, and universal access to economic resources as the foundation of the individual and collective good life in the United States, the current state ideology sanctifies the personal labor of reproducing life at work, at home, and in communities. That is, income-producing labor is deemed valuable chiefly in the context of its part in making smaller-scale, face-to-face publics.[29]

In other words, face-to-face and other sorts of emotional encounters that we might attribute to our biologically grounded intuitions must always be considered in a larger political context where the status of social institutions is at stake. And this cautionary tale supported by Darwin's *Expression*, I believe, applies equally to scholars in the humanities who might very well reject the politics of compassionate conservatism. No matter what the desired outcome, a model that squeezes culture between the pincers of a brain-world divide hampers our facility in cultural work and criticism.

In contrast, Darwin's rhetoric is essential to his nonreductive scientific project because emotions are themselves fundamentally rhetorical, which is to say Darwin's methodology accounts for the emotion's medium, occasion, and social situation. Consider again the word-image relationship, which was of crucial importance in Darwin's study of emotion and remains so to this day. Early in the *Expression*, Darwin describes his methodology while introducing an important caveat now typically sidelined: persuasive examples notwithstanding (for example, our Ekman exercise above), Darwin does not assume word-image identity as he recognizes how interpretation shapes each point in the equation.

Upon viewing the now famous Duchenne photographs, "several of the expressions were instantly recognized by almost everyone," notes Darwin, while others were subject to the most widely different judgments and pronouncements. But rather than explaining away this divergence like Ekman by way of an emotional hierarchy, "display rules" (*E*, 383), or stipulative definition, Darwin cautiously introduces the role of imagination:

> This exhibition was of use in another way, by convincing me *how easily we may be misguided by our imagination*; for when I first looked through Dr. Duchenne's photographs, reading at the same time the text, and thus learning what was intended, I was struck with admiration at the truthfulness of all, with only a few exceptions. Nevertheless, if I had examined them without any explanation, no doubt I should have been as much perplexed, in some cases, as other persons have been. (*E*, 21; emphasis added)

Here Darwin draws positive attention to the interpretive relationship between word and image, rather than justifying a reduction. Among other things this means drawing attention to the medium itself, which in this case is a set of photographs. Or more precisely, it is a report and discussion of an experiment designed around a set of photographs. And what about that photograph of horror as it appears in the book? It does not display horror as the woodcut might since, in this case, horror is obviously extorted by way of galvanism, and thus the photograph draws attention to its own staging. But don't all of these images, including those in the described experiments, stage emotion rather than instantiating some unmediated, face-to-face emotional event of the sort considered paradigmatic by Ekman? At the same time that Darwin exploits the reader's immediate reaction to an orangutan woodcut or galvanized horror, he foregrounds by way of narrative the emotion's mediation and thereby brackets the claim to some pure, face-to-face emotional experience.

This digression into imagination also allows Darwin to discuss emotions Ekman leaves off of his basic list, including love, sympathy, hatred, suspicion, envy, jealousy, avarice, revenge, deceit, devotion, slyness, guilt, vanity, conceit, ambition, pride, humility, and so on. Indeed the antiquation we may find in Darwin's list, where deceit and devotion are considered emotions, would suggest that human and not just evolutionary history is essential to Darwin's science. Witness Darwin on devotion:

> A humble kneeling posture, with the hands upturned and palms joined appears to us, from long habit, a gesture so appropriate to devotion, that it might be thought to be innate; but I have not met with any evidence to this effect with the various extra-European races of mankind. During the classical period of Roman history it does not appear, as I hear from an excellent classic[ist], that the hands were thus joined during prayer. Mr. Hensleigh Wedgwood has apparently given the true explanation, though this implies that the attitude is one of slavish subjection. "When the suppliant kneels and holds up his hands with the palms joined, he represents a captive who proves the completeness of his submission by offering up his hands to be bound by the victor. It is the pictorial representation of the Latin *dare manus*, to signify submission." Hence it is not probable that either the uplifting of the eyes or the joining of the open hands, under the influence of devotional feelings, are innate or truly expressive actions; and this could hardly have been expected, for it is very doubtful whether *feelings, such as we should now rank as devotional*, affected the hearts of men, whilst they remained during past ages in an uncivilized condition. (*E*, 217–218; emphasis added)

Civilization entails, apparently, a distinct emotional regime that separates it from the uncivilized, thereby contradicting an evolutionary theory that sees our continuity with Pleistocene hunter-gatherers but doesn't see the distortions inevitably imposed by our interpretive lens. According to Darwin we need historical narratives (however dubious in this instance) first to denaturalize what might appear an innate feeling such as devotion, and second to explain in positive terms a feeling that we might experience and observe as universal. This is Darwin's skeptical work. Darwin's historical work is part and parcel of his careful scientific observation; without it the "we" of his first audience would naturalize some particular posture and thereby see the emotional phenomenon poorly. Meanwhile the "we" of his universal audience fails to recognize a particular posture at all, which is itself crucial data revealed by way of comparative anthropology and human history. In his microstudy of devotion and other emotions, Darwin takes into account the emotion's linguistic and visual mediation, its occasion in the service of a pointed argument, and personal perspectives of varying historical immediacy. That is to say, again, that Darwin's scientific theory of emotion is inherently rhetorical, which is a methodological strength instead of a weakness. Of course, Ekman would ignore a phenomenon like devotion altogether.

And what about love? We can return to Darwin in our effort to expose the sleight of hand where Ekman offers the feeling of disgust as the model for all emotions while elsewhere he defines "all emotions" so narrowly that the list excludes love and hate amongst others. "No emotion is stronger than maternal love," qualifies Darwin in a Victorian mood, "but a mother may feel deepest love for her helpless infant, and yet not show it any outward sign" (*E*, 82). Indeed, Darwin doubts whether a long list of what he calls complex states of mind are revealed by any fixed expression and are sufficiently distinct to be described or delineated in strictly scientific terms. But unlike Ekman, Darwin turns to the literary humanities for a demonstration of how any science of emotion must ultimately make room for interpretation. When "Shakespeare speaks of envy as '*lean-faced in her loathsome ease*'; or *black*, or *pale*, and jealousy as '*the green-eyed monster*'; and when Spenser describes suspicion as '*foul ill-favoured, and grim*,' they must have felt this difficulty. Nevertheless, the above feelings—at least many of them—can be detected by the eye; for instance, conceit; but we are often guided in a much greater degree than we suppose by our previous knowledge of the persons or circumstances" (*E*, 260).

Ekman recognizes this complication but circumvents the implications by relabeling everything beyond his seven basic emotions "affective commitments." "Emotions are brief and episodic, lasting seconds or minutes" explains Ekman. "Parental love, romantic love, hatred, envy or jealousy last for much longer periods—months, years, a lifetime for love and hatred, and at least hours or days for envy and jealousy. In each of these there is a strong commitment, an important attachment (negatively in the case of hatred, envy and jealousy). They also differ from emotions in specifying something about the cast of characters involved" ("Commentaries," 83).

Of course, in noting how the cast of characters is essential to the composition of supposed nonemotions including love, hatred, envy, and jealousy, Ekman seems to invoke Aristotle's famous discussion of the emotions in book two of the *Rhetoric* just before he breaks off abruptly. Darwin, by contrast, writes long and fascinating Aristotelian passages where the cast of characters that compose an emotion are specified in their complex sociopolitical relations. Take, for instance, Darwin on hatred:

> If we have suffered or expect to suffer some wilful injury from a man, or if he is in any way offensive to us, we dislike him; and dislike easily rises into hatred. Such feelings, if experienced in a moderate degree, are not clearly expressed by any movement of the body or features, excepting perhaps by a certain gravity of behaviour, or by some ill-temper. Few individuals, however, can long reflect about a hated person, without feeling and exhibiting signs of indignation or rage. But if the offending person be quite insignificant, we experience merely disdain or contempt. If, on the other hand, he is all-powerful, then hatred passes into terror, as when a slave thinks about a cruel master, or a savage about a bloodthirsty malignant deity. (*E*, 234)

So in this case and others like it the very experience of an emotion, according to Darwin, depends upon the social and not the biological situation, namely the relative significance of the characters involved—including the subjective perception of such—which can be systematically distorted, of course, by hegemonic constructs including racism and patriarchy.[30] By way of grammatically loaded micronarratives Darwin, like Aristotle, takes into account the uneven social relationships that instantiate an occasion for hatred (having suffered), the phenomenological time of hatred (expecting to suffer), and the intentionality of hatred (as a function of willful injury not accident). There is no way a facial expression or isolated brain scan can adequately represent an emotion like hatred, and simply defining it like Ekman as nonemotional begs the question.

Finally, when Darwin wishes to discuss these so-called inactive emotions he again finds recourse in the visual arts and literature, and not just as a repository for accidental wisdom in the manner of Damasio, who also likes to quote Shakespeare, but rather as a necessary component of a larger theory. Consider the famous first-century sculpture (Fig. 1.6) where Laocoön has been variously considered "bellowing" like a bull (*mugitus*) by Virgil,[31] "noble" (*edle*) by the philologist Winckelmann,[32] "groaning" by Lessing,[33] and silently struggling by Charles Bell, whose book *The Anatomy and Philosophy of Expression as Connected with the Fine Arts* served as Darwin's polemical foil.[34] In Bell's *Anatomy*, Darwin found a sketch (Fig. 1.7) that he relies on to invoke Lessing's famous argument that the visual arts must disperse violent emotions. Visual arts provide an insufficient venue for the scientific study of emotion, warns Darwin, because, just as Lessing suggested, contracted facial muscles destroy beauty.

Insufficient, yes, but still necessary in a rich research program where one is obligated to analyze emotion beyond measurable facial expressions. As in literary narrative, where an emotion such as hatred can emerge only through

FIGURE 1.6. "Laocoön and Sons." (Photograph by Livio Andronico, CC BY-SA 4.0)

a cast of characters, in the visual arts "the story of the composition is generally told with wonderful force and truth by skillfully given accessories" both human and nonhuman, animate and inanimate, foregrounded and backgrounded (*E*, 22).[35] In this case, the skillfully given accessory of the sea serpent is crucial for any justifiable interpretation of the emotion represented, for we are not talking about orgasmic ecstasy, after all.[36] Other key accessories that tell this emotional tale "truthfully" include Laocoön's two sons perishing beside him, who thereby force the scene beyond the emotions of noble self-sacrifice that otherwise might look resolute. And ultimately these immediate characters and accessories offered up to the eye emerge only against the receding background of mythical characters that include the vengeful Apollo, who is punishing his disobedient priest, and the Trojan people for whom Laocoön sacrificed himself by warning them of gift-bearing Greeks.

So where is the emotion (Fig. 1.8)? Is it simply in that first face as the Ekman experiment insists? I offer the joystick in case you are thrown by the anthropomorphism of a smile on the middle image, but in fact I think the

McDonald's Happy Meal provides an efficient countermodel to Ekman's pseudo-Darwinian program. After all, the Happy Meal is much more than a burger, fries, drink, and plastic toy, more than simply an emotionally competent object (like Ekman's delicacy) triggering happy faces like the one next to it. Rather, the Happy Meal is an emotional artifact seemingly untethered to—Karl Marx would say alienated from—the complex social transactions that produce it. So instead of turning toward Ekman as the supposed champion in our day, I heretically prefer to imagine Darwin turning toward his contemporary Marx as he says in this case an accessory that tells the emotional tale has apparently taken on a life of its own. I think revisiting Darwin's

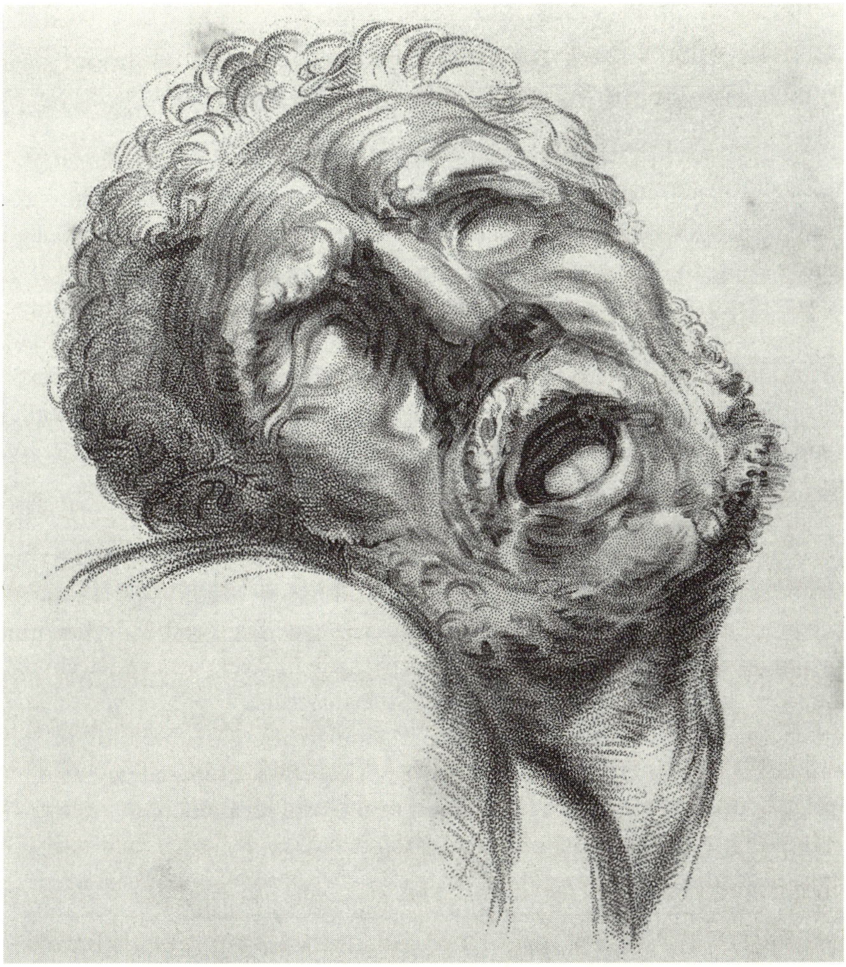

FIGURE 1.7. Sir Charles Bell, *The Anatomy and Philosophy*. (Wellcome Library, London, CC BY 4.0)

FIGURE 1.8. [Left] Photograph 1 in a sequence of smile intensities (PF1–6) from the Pictures of Facial Affect (POFA) Collection. (Reproduced with permission from Paul Ekman Group, LLC.) [Center] General Happy Meal Box. [Right] "Joystick black red petri." (Digital image by Ansgar Hellwig, PhD)

work on emotion suggests this turn to sociorhetorical analysis is at least as plausible as the other.

In his foreword to Andy Clark's *Supersizing the Mind: Embodiment, Action, and Cognitive Extension,* David Chalmers suggests we might also research extended desires, extended reasoning, extended perception, extended imagination, and "extended emotions."[37] "One might have something akin to an extended mood, if not an extended emotion, when one's environment is always nudging one toward happiness or sadness," offers Chalmers by way of example,[38] and certain Extended Mind (EM) theorists such as the philosopher of biology Alva Noë have begun to work in this direction. "In the Beginning Was the Situation," offers Noë, not some Cartesian computing machine that builds up internal pictures of the scene, makes plans, and executes them (*Out of Our Heads,* 101). Moreover, according to Noë, these "situations" are profoundly characterized by emotion. If, for instance, a psychologist is trying to understand why an infant will become distressed when his or her mother becomes still or impassive, that psychologist would go astray in positing the infant as a separate individual observing the mother. Instead we should consider the mother as "literally one of the structures constituting a child's psychological landscape": the baby's relation to the other unfolds in "an emotional setting" (*Out of Our Heads,* 31). Likewise, Hubert Dreyfus—an influential EM theorist in the tradition of philosophical phenomenology—argues that any artificial intelligence project designed to approach humanness must account for emotional settings, and to do so it "would also need . . . a model of *our particular way of being embedded and embodied* such that what we experience is significant for us in the particular way that it is. That is, we would have to include in our program a model of a body very much like ours with our needs, desires, pleasures, pains, ways of moving, cultural background, etc."[39] But for all this interest in emotional and cognitive settings and situations, EM theorists like Dreyfus, Chalmers, Clark, Wheeler, Thompson, and Noë come up short when describing this worldly side of the brain-body-world nexus. To put it bluntly, they too need the arts and literature in the spirit of Darwin. Since sentimental literature is central to the Cognitive Approaches movement, I conclude this chapter with an example from Sterne's *A Sentimental Journey through France and Italy*, which is a masterful study of emotional situations of a certain sort.

As Yorick travels through the treacherous mountains between Turin and Savoy, he reports how a journey that is potentially horrible fares better with unhurried sentiments giving way to this observation of peasants encountered *en route*:

> Poor, patient, quiet, honest people! fear not: your poverty, the treasury of your simple virtues, will not be envied you by the world, nor will your valleys be invaded by it.—Nature! in the midst of thy disorders, thou art still friendly to the scantiness thou hast created—with all thy great works about thee, little hast thou left to give, either to the scythe or to the sickle—but to that little thou grantest safety and protection; and sweet are the dwellings which stand so shelter'd.[40] (120)

Nowhere in this scene do we find suggestive documentation of face-to-face encounters à la Paul Ekman. In fact the pathos of this scene thrives on a variety of distancing mechanisms familiar to the landscape artist and the critical theorist, respectively.

At the very least Yorick's pity for the peasants must be understood in contrast to the geographical horror it supersedes, thereby escaping a Steven Pinker type of literary Darwinism where the novel is "cheesecake for the mind," and novelistic fear provides an opportunity for virtual experience of a situation better avoided in real life.[41] Immediately following upon a "feast of love" at the table of a large and respectful French peasant family (119), this scene of *partial* spectatorship restages the classically tragic sequence that moves from horror to pity as a critical opportunity (as we will see in a minute), not as a virtual experience.[42] Yorick's pitiful perspective can be understood only against a romanticized landscape that includes—along with sweet dwellings—poor, patient, quiet, honest people who would no doubt have a very different perspective on the situation, if asked. And I imagine most of this situational complexity, including its class element, is not lost on the ingenious author, who earlier outlines the terms of a "sentimental commerce" where instead of natural immediacy, sentiment is produced by the imbalance of social exchange (9). Narrating the start of his sentimental journey, Yorick explains that being-out-of-place produces an imbalance of sentimental commerce where conversation, for instance, comes at a high price: think humiliation over a botched commercial transaction or awkwardness when attempting a joke in a foreign tongue.

Certainly knowing "whose sentiment it is" constitutes a crucial aspect of our understanding of the psychological dynamics of this particular scene and the novel as a whole, as literary critic Lisa Zunshine might insist.[43] But if, like Zunshine and many of her cognitivist colleagues, we assume that our own reading mind and the "mind behind the sentiment" function like a Cartesian computing machine managing information and generating metarepresentations, then we desensitize ourselves to the relevant cultural institutions of language and perception, and a scene like this can be reduced to "imagined landscapes with their pathetic fallacies, personifications, and anthropomorphizing" (27, 47, 63). And contra the critic of sentimental literature June Howard (quoting Sanchez-Eppler), these words on the page do not produce in the reader "pulse beats and sobs" that radically contract the distance between narrated events and the moment of their reading (though some of Sterne's racier moments admittedly do, for this reader at least). Quite the opposite—by way of negation, Sterne stages the empathic moment with his typical irony that suggests we take no emotion at face value.

Fear *not*, envied *not*. What do negated emotions feel like after all? Look like? What bit of information about the real world do they represent? Where might they be located on the face, or mapped in the brain? It turns out that the most emotionally pregnant words in the passage—*fear* and *envy*—appear only in negation, which blocks the royal road to experience we might very well expect on a sentimental journey. The straightforward command "Poor, patient, quiet, honest people! *rejoice* . . ." would have been the more immediate stylistic solution Sterne avoids in favor of his negation that invokes a particular "world" at the same time that it forces readers to account for multiple perspectives, including our own, as we map the emotional terrain. That's one function of negation as we will see in Austen as well: the rhetorical situation including the reader is, in this case, implied.

Why the negative at all, let alone the double negative, when a logician like Bertrand Russell insisted the world can be described without the use of the word "not"? Clearly in Yorick's case the impotent command "fear not" differs from both the command "rejoice" and the condition of fear*less*ness, which would exploit the situation differently by way of an evaluative judgment (no one calls himself fearless, after all).[44] "Poor people," "scythe" or "sickle," "sweet dwellings," "valleys," "an unenvious world." Emotion maps these relationships—character, tool, dwelling, landscape, world—always considering the dynamic frame itself as a factor in the equation that renders the world *un*envious, implying that we should not take the world for granted as an unarticulated background, or as a natural environment, or as a comfortable setting-in-general, or as a comfortable situation. In short negation is not just a logical operator. Negation has a rhetorical function.

To explain how this is so, I offer a brief excursus on Laurence Horn's neo-Gricean masterpiece *A Natural History of Negation* as it distinguishes amongst negation's numerous functions—including logical—which is *not* reflected by a comparable symmetry in language structure and language use (xiii).[45] Horn's indispensable contribution comes when he pays close attention to this asymmetry across instances that would typically escape philosophers and linguists like him, while offering analytic continuity with literary examples taken on their own terms, as I will show using this example from Sterne, and Austen.

What should we do without double negatives in Jane Austen's *Emma*, for instance, where Harriet can stand in "no unhappy reverie" completely different from the logically equivalent "happy reverie" that would lose the implication of surrounding characters, narrator, author, and reader? Here's the passage at issue: "Harriet, who had been standing in no unhappy reverie, was yet very glad to be called from it, by the now encouraging manner of such a judge, and such a friend as Miss Woodhouse, and only wanted invitation, to give the history of her hopes with great, though trembling delight."[46] Though this example from *Emma* isn't his, Horn would no doubt draw our attention to the traditionally suspect rhetorical figure litotes, which is a form of understatement where an affirmative is expressed by the negative of the contrary (303). What is the function of litotes? Typically Horn sets the stage with some natural history as he calls it. Just as hyperbole (according to Aristotle's *Rhetoric*) is a figure fit for young men of quality, and ellipsis the favored figure for politicians, Alexander Pope under a pseudonym sees in litotes "the peculiar Talent . . . of Ladies, Whisperers, and Backbiters" (quoted in Horn, 304). Instead of logically dissolving litotes or condemning it as inherently seditious, however, Horn asks us to consider the crucial implications when stereotypic situations are pragmatically avoided. Horn suggests "the corresponding periphrastic forms, stylistically less natural, longer, and more complex, are restricted, via Q-based implicature to those situations *outside the stereotype*" (304). That is to say Horn invokes the Maxim of Manner according to Grice: be perspicuous, avoid obscurity and ambiguity, and strive for brevity and order. So when violated with something like litotes, the Maxim of Manner suggests you are getting at something *else* altogether. And I would concur that Austen's double negatives function just this way, as they render critical situations that refer to, but exceed a stereotype: in this case a happy reverie that would leave out the "judge" and "friend" Emma Woodhouse as well as the narrator who suggests indirectly the two descriptions are mutually exclusive. Since when, after all, is a judge your friend or vice versa? Not recently, I would imagine, but to get this point and to understand how it precipitates no unhappy reverie on the part of Harriet, "cultural background etc." needs to be rendered in detail that exceeds the capacity of philosophical pragmatics narrowly conceived.

The emotional phenomenon does depend upon background—but we must distinguish what this means contra pragmatic philosophers. We are not talking about the unarticulated "background of intentionality" à la John Searle, or differently Hubert Dreyfus, which set the stage inadequately if we use Horn's model to look at examples from literature.[47] Dreyfus is right that the satisfactory gestalt that causes an action is in no way represented when it comes to an expert tennis swing or a master chess move, for example (29, 32). Likewise literature does not represent a satisfactory gestalt that causes an action even when it can avail itself of copious description, or a narrator who seems to tell us exactly what causes an action like Harriet standing—namely a reverie. But the quality of nonrepresentation across these examples differs significantly. In the examples from Dreyfus, nonrepresentation is an afterthought when we comport ourselves normally. In the example from Austen as interpretable through Horn, nonrepresentation is the result of *flouting* a conversational norm, which is something we do all the time. And Horn might agree that literature—not everyday activity—is especially suited for this kind of abnormality because it takes definitive shape in unnatural styles, unnecessary length, and undue complexity.

Or to put this in positive terms, literature is in a special position to teach us about normality and its construction. Literature and especially certain kinds of novels provide the richest information for reading rhetorical situations dynamically, with framers such as the reader implicated sometimes beyond what might be imagined by an author who is not a feminist in the case of Austen, and not a Marxist in the case of Sterne.

Returning to our main line of inquiry, we now want to know where in Sterne's material "cognition" is located and how it works. Some might say cognition is embodied and embedded in the world, composed in particular (but not infinitely variable) ways. But against the grain, Yorick's "world" does not envy the poor, it disregards the simple virtues, it avoids the valley as does Yorick himself—again the work of negation. So this passage carefully characterizes Yorick's "cognitive world" as Wheeler might put it, which Yorick shares partially with certain others like the attractive lady he next encounters (121), but less so with the poor of any sort. Remember, however, that Yorick explicitly disavows such characterization—a world of brutal gentrification after all—instead throwing in his lot with friendly "nature," before passing by the scene untouched, relieved of the geographical terror he faced on high. Hypocrisy yes, but that's not the end of it. Cognition in this scene is never comfortably embedded in the scenery (as some theorists might imagine) nor is it detached (as other theorists might imagine): it is practically situated but *uncomfortably* so, which points us to another order of the emotional phenomenon where the author and reader appear on the scene. Another way to put this: the terrible discomfort of the character Yorick is pseudo-resolved diegetically, in telling the story from a first-person perspective, which in turn produces second-order discomfort in a reader who alone can register the irony of this scene that beautifully "models" a social world where class makes all the difference down to the most minute details of perception, action, emotion, and cognition; remember how the story might be told differently from the peasant's perspective. Such is the critical moment essential for reading sentimental literature, including less ironic works like Fielding's *David Simple* or Henry McKenzie's *The Man of Feeling*, which nonetheless have a critical edge insofar as they each implicate a world turned upside down by men who feel.[48]

Working against the Ekman trajectory in cognitive science, I argue that an emotional phenomenon, including love or a so-called basic emotional phenomenon like terror, does not announce itself unequivocally. As Griffiths and Scarantino argue, indeterminacy must be in-built for the phenomenon to emerge as emotional in the first place, and sentimental literature provides sophisticated models for how this indeterminacy takes shape in forms we call historical.

The mind is never comfortably embedded in the scenery (*pace* EM theory) nor is it detached (*pace* certain forms of CAL): it is practically situated but uncomfortably so, which points us to another order of the emotional phenomenon where the audience appears on the scene. This is the critical moment, essential for reading literature as we will see in subsequent chapters, and for reading faces, as Darwin insisted. Emotional phenomena do not announce themselves unequivocally. Indeterminacy (but not radical indeterminacy) must be in-built for the phenomenon to emerge as emotional in the first place.

What might this story look like from the perspective of evolutionary biology? Would it not be advantageous if we all wore our emotions on our sleeves, so to speak, plastered to our foreheads so anyone could read them without equivocation like the Ekman fantasy, or they were available by way of Vulcan mind meld, which would eradicate the need for emotions at all, as the creators of *Star Trek* so cleverly realized? Isn't emotional indeterminacy maladaptive, as evolutionary biologists might say, a species trait that should have been selected against over the long term in favor of more direct and ostensibly more practical communication? Though here I cannot make the case in detail, I can point out how Stanford biologist Joan Roughgarden and others working on biodiversity provide an alternative perspective more amenable to our skeptical Darwin and the revised consilience he can offer. Instead of selecting sexually for ideal types, argues Roughgarden, a species needs "a balanced portfolio" of genes to survive over the long term,[49] and sex, which entails a very wide (but not indefinite)[50] range of behaviors—reproductive and otherwise—is the *social* activity that continually rebalances a species' overall genetic portfolio in the context of dynamic environments. Instead of offering only background noise, indeterminacy of the sign (as we might call it from the semiotic perspective, where X is somewhere between attractive or repellent, pro- or antisocial, praise or blameworthy, and so on), is compatible with biodiversity insofar as it constitutes the social. Antisocial eugenics or cloning are Roughgarden's counterexamples; just like the computer scientist knows that focusing only on the code while ignoring the execution environment is a mistake, cloning biologists who focus on the nucleus of the cell while ignoring the cytoplasm make the same mistake insofar as they have ceased to work ecologically.[51]

Consider, finally, one of Darwin's favorite subjects in the *Expression*, the infant displaying "mental distress" (Fig. 1.9).[52] He might be hungry, sleepy, gassy. Perhaps there are a few more possibilities but not an infinite number, or at least that's what we like to think. But wouldn't it be more practical, more adaptive, if the label, "hungry!" appeared on the forehead of an emotional infant or if we could reliably read the infant's mind? Why aren't we permanently sequestered in utero so that all of these communication problems might be avoided in the first place? In the spirit of Roughgarden we might observe the semiotic indeterminacy of emotional expression is a corollate of biodiversity precisely because it necessitates communication; the work of triangulating is never done nor should it be.

We need the arts and literature to study emotion because they transpire on the appropriate phenomenological level where experiences happen, on the one hand, and where experience is characterized, on the other. Just as we cannot understand bird flight by studying only feathers, to cite Noë's example, we cannot understand emotional experience by studying only the face, the eyeball, the ear, the brain (see *Out of Our Heads*, 157). To understand

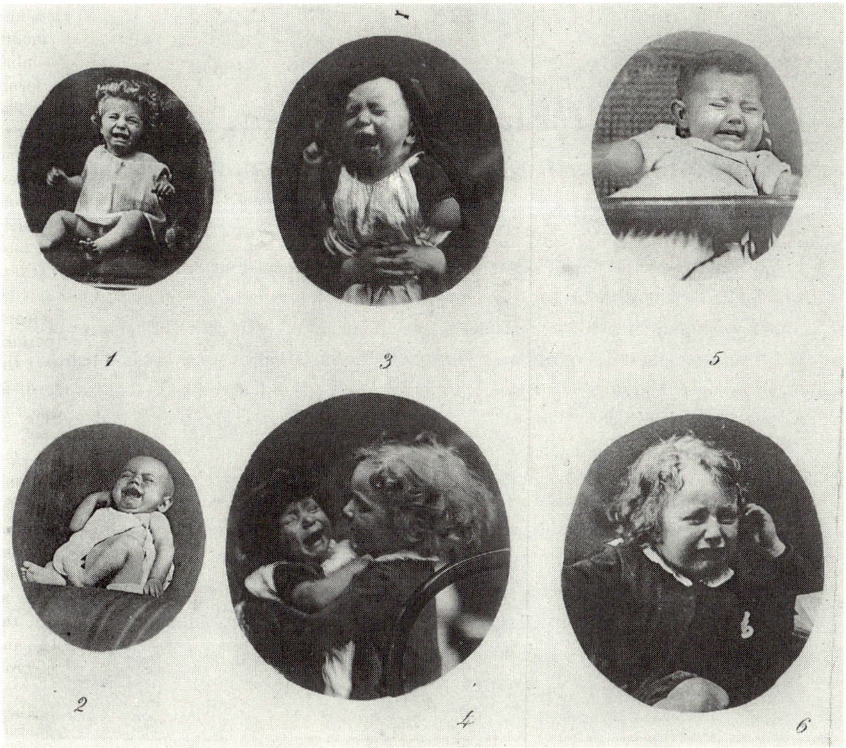

FIGURE 1.9. "Weeping Children from Darwin's *Expression of Emotions*." (Wellcome Library, London, CC BY 4.0)

experience beyond having it we need human beings in the world characterized, which is to say we need people in situations. And not just people in situations isolated for the purpose of answering isolated questions in the mode of social scientific inquiry, though this approach also has its utility. Our people in situations should be subject to the maximum degree of manipulation and critical scrutiny, reflection from inside and outside, and subjected to all sorts of imaginative possibilities so that even the complexities of experience that shimmer just beyond our immediate grasp still can make their presence felt by way of tropes, like irony, that help define the humanities. As Darwin understood over a century ago we need the arts and humanities for many reasons, including for the sake of good science that can account for both the basic ways we are, and the ways we can be.

2

Bearing Up in *The Interesting Narrative of the Life of Olaudah Equiano, or Gustavus Vassa, the African*

> It was now between three and four years since I first came to England, a great part of which I had spent at sea; so that I became inured to that service, and began to consider myself as happily situated; for my master treated me always extremely well; and my attachment and gratitude to him were very great. From the various scenes I had beheld on ship-board, I soon grew a stranger to terror of every kind, and was, in that respect at least, almost an Englishman.
>
> EQUIANO, *Interesting Narrative*[1]

> Let us therefore say . . . that the life of consciousness—cognitive life, the life of desire or perceptual life—is so tended by an "intentional arc" which projects roundabout as our past, our future, our human setting, our physical, ideological and moral situation, or rather which results in our being situated in all of these respects. It is this intentional arc which brings about the unity of the senses, of intelligence, of sensibility and motility. And it is this which "goes limp" in illness.
>
> MERLEAU-PONTY, *Phenomenology of Perception*[2]

Having read this far, you can guess I'll make use of Merleau-Ponty's going limp, his illness that speaks to a world turned upside down. But first we get Merleau-Ponty's rigor and his sense of stability, which is also where we begin reading Equiano's *Interesting Narrative*—that is, somewhere in the middle when things are bearing up. What would it mean to be "happily situated" as a slave? What things appear within reach at this point in Equiano's narrative? On first impression it would seem Equiano provides an answer that slips conveniently into our sciences of happiness—notably Daniel Kahneman et al.—and our theories of situated cognition drawing directly from Merleau-Ponty, as in the notable work of Alva Noë, including *Out of Our Heads: Why You Are Not Your Brain, and Other Lessons from the Biology of Consciousness* (2009). In short, Equiano adjusts to his situation. But are we sure this is the story? My demonstration in this chapter proceeds through Equiano's (1) time, (2) attachment/feeling, and (3) bearing up, while remaining attentive to the profound inconveniences that Equiano suffers despite whatever comfortable accommodations appear on the horizon. Slavery as inconvenient—a rather absurd understatement—actually points to the analytic problem of this chapter. In each of these three positive instances—time, attachment, bearing up—I'll show how Equiano's testimony speaks loudly to our sciences of happiness and situated cognition, before dissipating into queries and quibbles and counterarguments that should, in the end, threaten any easy sense of well-being. Discomfort remains profound. I will thus be working a local problematic at the intersection of science and the humanities, paying special attention to rhetorical qualities of a phenomenon—well-being—that escape our scientific interests, narrowly conceived. At the same time, with Equiano's help, I'll attend to hedonic (happiness) psychology and theories of situated cognition where resilient threads from the humanities might be reinforced. Both of these science-oriented projects, it turns out, have an impulse that is explicitly rhetorical, which means they wind up sharing significant ground with the literary humanities.

Time Doesn't Add Up

The first object that "salutes" young Equiano's eyes when he arrives on the African coast after capture—a slave ship—resets his calendar to year zero. It marks a terrible rebirth, or renaturing where nothing is yet known or properly incorporated; at the same time, it provides an initial reference point for experiences that will now accrue in some new and more normal fashion. Following upon zero, the ship fills Equiano with "astonishment, which was soon converted into terror" (55) because it was something he had never experienced (56). Equiano is at that point more persuaded than ever that he was "in another world, and that every thing about . . . was magic" (59), recounting, in excruciating detail, how in this magical new world even familiarities like fish begin to fly, clouds appear as land through the mariner's quadrant, and the familiarities of basic sense decompose. The terrible closeness of place, the heat of the climate, the copious perspiration, the loathsome smells, the groans of the dying, the galling of the chains, the altogether "wretchedness of the situation" on a slave ship undoes completely, just as it was designed. "Insupportable" names for Equiano the living conditions that collapse around him, while "inconceivable" names the conditions for knowing that collapse at the same time (58). "Every object was new to me," recounts Equiano,[3] referring both grammatically to houses and horses he sees upon landing in Barbados, and also narratively, to a merchant's yard where the live cargo, including himself, were all "pent up together like so many sheep in the fold, without regard to sex or age." Here, in this new world, "every new object" includes familiars like mothers and children who now appear totally different, like so many sheep (60).

So many sheep. Equiano's year zero foreshadows, at the same time, just what you might imagine, a series of seconds where the calendar would be reset once again: second nature, second coming. Equiano's tortured rebirth into a world turned upside down, and even his appearance as one of so many sheep, suggests that the signs of this new world point in all their perversity toward the next world, which seems to always shimmer in the background even when it isn't immediately recognized as such. "I was from early years a predestinarian," recalls Equiano. "I thought whatever fate had determined must ever come to pass" (119); which leaves the reader to wonder how early, exactly? Evidently before even his exposure to Christianity proper. Within the first handful of pages, Equiano recalls Igbo native beliefs: "there is one Creator of all things . . . who governs events, especially our deaths or captivity" (40). And it is no accident this (same?) Creator "who in very deed leads the blind in a way they know not" soon after appears to his comfort, even before Equiano's conversion to Christianity appears in the narrative proper (63). Remember, "comfort" in the eighteenth century describes first what God can give you. Comfort is only secondarily what the world can give you, and it tends to arrive "overrated" as creature comfort. (In 1769, as the OED documents, B. Wallin writes in the *Eternal Mansion*, 31: "a just estimate of this life is becoming and useful; but we are prone to over-rate creature-comforts, and place too much upon them.") Equiano's interesting narrative certainly does mark a conventional chronology as we learn about his life early-to-late, including his by-the-book rebirth into Methodism, which Merleau-Ponty might identify as the "intentional arc" of religious experience. And which, we will see, recapitulates the rebirth he suffers into slavery when Equiano is resurrected from social death[4]: his senses are predictably renewed—touch, taste, smell, vision—Merleau-Ponty's "unity of the senses" finally achieved. No doubt his rebirth does provide some spiritual comfort in the face of creaturely comforts withheld, at least for the time being.

However at the same time, each moment, each experience, fractures chronological time into arcs that point elsewhere and unpredictably, without ever achieving a godlike comprehension that would settle everything. Such perspective only God has, after all, so a mere mortal like Equiano can only read the signs, which are diverse and don't necessarily add up. Senses of the old creature fail to guide practically: the blind know not where they are led (Isaiah 42:16). And sacred rhetoric likewise reverses the temporal order of call and response, the sensual order of the voice before the ear: "Before they call, I will answer; and while they are speaking, I will hear" (Isaiah 65:24, quoted in *Interesting Narrative*, 182). Unity of the senses would thus find its organ *outside* of any particular creature as such, organized into patterns that predate and antedate creaturely life. After capture, what Equiano "sees" and "tastes" in an African town called Tinmah (52) clocks as a natural resource—sugar—that might in the future be profitably exploited (i.e., cultivated in Africa by free native labor, Vincent Carretta notes, 52n95). So Equiano sees the coming world of free commerce, which would no longer rely upon slave labor. At the same time, with the help of John Gill amongst others, Equiano sees in the manners and customs of his countrymen not just Africans, but Jews "in the pastoral state," before they reached the Land of Promise (43), but after the world has fallen according to Genesis. To summarize, following Equiano we find at least a) ontogenic time, which sees the world of Equiano's narrative begin at the beginning, "I was born . . ."; b) historical time, which sees the commercial world sometime after 1794 when the ninth edition of the *Interesting Narrative* was published still in Equiano's lifetime; c) prophetic time, which sees the world coming to an end; and d) the time of predestination, or perhaps, more accurately, "prevenient grace," which precedes human decision, ordering my (perceptual) world thus not so, as it cuts through the other three temporalities ontogenic, historical, and prophetic.

"Almost an Englishman" has pointed interpreters of Equiano's split sense to W. E. B. Du Bois on "double consciousness" of a black person in a predominantly white society. Among others Vincent Carretta and Henry Louis Gates Jr. give us this influential and not inaccurate line of interpretation.[5] But at the same time I hope my interpretation of Equiano's sense does not split in two, if that means each perspective can be measured against the other thanks to some common denominator. In *The Souls of Black Folks* (1903) we read: "It is a peculiar sensation, this double-consciousness, this sense of always looking at one's self through the eyes of others, of measuring one's soul by the tape of a world that looks on in amused contempt and pity" (3). No doubt Du Bois provides here one important challenge to Kant's unity of experience enjoyed by consciousness that cannot split in principle because it is removed from something particular like a black body.[6] Following Du Bois, the doubling of consciousness decenters experience so that its rightful ownership is called into question along Hegelian lines.[7] If one "looks through the eyes of others," then what one perceives, or experiences, will not be one's own in some absolute sense. We are obliged to a critical investigation of rightful ownership in principle and in practice, where in each case the question is: By what power does sense experience take shape? Some responses follow.

Writing in 1952 about "The Lived Experience of the Black Man," Frantz Fanon starts with a brutal observation from the perspective of white society: "'dirty nigger!' Or simply 'Look! A Negro!'" (89). In a section on "The Black Man and Hegel," Fanon concludes for the black Frenchman, "the situation is unbearable" (196), and I will return to this formula in the last section of this chapter on bearing up.[8] In 1990 Iris Marion Young addressed this question—By what power does sense experience take shape?—from a feminist perspective, with the help of Merleau-Ponty and Simone de Beauvoir: "women in sexist society are physically handicapped" (42). They tend not to put their whole bodies into engagement in a physical task with the same ease and naturalness as men; the space available is constricted (33). On this model "sense data" is not neutral as it takes shape in the context of patriarchy; the data available to a monitor placed on the skin is, in this case, determined by the social environment as is physical capacity—hence the title of Young's famous 1980 essay, "Throwing like a Girl."[9] Likewise in *Assuming a Body: Transgender and Rhetorics of Materiality*, Gayle Salamon has recently asked, "What are the consequences for embodiment and subjectivity when the relation between self and skin is not one of ease and euphoria but discomfort and dysphoria" circuiting through the gaze of an external world "hostile to gender ambiguity"? (27).[10] After Hegel and Du Bois, much has been made of the power that divides consciousness, as it seems to locate sense in the organs of the other: "My my look at you!" Tempting is a rejoinder that reasserts the truth of self-consciousness, hoping that I, or those we care about, will somehow measure up.

But what do we make of Equiano himself, as he becomes "almost" (but not quite) an Englishman, and thus would never be able to measure up from any perspective, let alone from the perspective of God who appears on the scene consequentially? As the Old Creature, remember, Equiano's senses fail him in every way possible and rehabilitation or even reparation can never appear fully on the horizon. At just this point Fred Moten pursues Equiano beyond double consciousness: "What I'm after," explains Moten, "is a kind of knowledge that moves from somewhere on the other side of either reason or experience, intelligibility or sensibility, and that is not reducible to any originary state of nature but for that improvisation of the human which is neither the encoding of or embeddedness in responsibility" (299).[11] And yes, I am also arguing that *embeddedness* in responsibility must be negated as we approach Equiano, because the terms of responsibility that would give him some comfort and some sense of normalcy—for example, the "gratitude" Equiano expresses toward his master—remain immeasurable and untenable.[12] After all, the slave master Equiano owes gratitude in this case is Michael Henry Pascal, who sells Equiano just when freedom seems within reach (93–94). How does one measure gratitude like that? At the very least with some measure of irony, which is how we must understand our opening passage composed by Equiano, of course, with full knowledge of the betrayal to come: "Thanks a lot."

This last temporality, prevenient, puts much of the "interesting" into Equiano's interesting narrative because it finds a place for the untimely. Prevenient grace makes it interesting that Equiano fell unhurt from the ship's ballast while fighting in the Seven Years' War, and it makes interesting other instances that might not otherwise fit into the narrative structure of the sea yarn, the travelogue, the picaresque saga, or even the standard conversion narrative.[13] Equiano appeals to the reader: "I beg leave to relate another instance or two which strongly raised my belief of the particular interposition of Heaven, and which might not otherwise have found a place here, from their significance" (87).[14] This narrative logic of prevenient grace does not trump ontogenic or historical or prophetic temporalities, but it does suggest all sorts of tangibles that turn out to be very interesting indeed. Equiano's justification is not titillation but simply what he senses, or more precisely, what *he thinks he senses* sketched out in the world: "In these, and in many more instances [e.g., cheating death], I thought I could very plainly trace the hand of God, without whose permission a sparrow cannot fall. I began to raise my fear from man to him alone, and to call daily on his holy name with fear and reverence" (88). Crucially this transition takes us from a narrative sensibility—*dear reader, notice this not that*—to an emotion, fear, as it is displaced from the narrative per se and reoriented toward God. More on this fear later as it makes room for Equiano's happy situation here on earth.

Nevertheless, Equiano happily situated just about ten pages prior recalls blandly the passage of calendar time: it was now between three and four years since he first came to England. And as our theories of situated cognition and our new sciences of happiness would have it, this mere passage of time should in fact make a real difference. In their work on "Norm Theory: Comparing Reality to Its Alternatives," for instance, Daniel Kahneman and Dale Miller discuss the phenomenon they call "emotional amplification," which states that the affective response to an event is enhanced, relative to the response to the same events when they are normal and expected. The special displeasure or pleasure derived from abnormality—say Helen has switched to a new cereal which she likes much better than the one she had consumed for a long time—will inevitably disappear as the norm gradually changes to reflect Helen's new routine.[15] Similarly over time, let us say with Kahneman, Equiano becomes inured to the services he provides as a slave, and a stranger to terrors of every kind. As Kahneman and Miller might analyze Equiano's situation at that point, the availability of counterfactuals "define[s] the grain of his experience" and so with the mere passage of years on the calendar, Equiano's expectation that he would return to his previous life in Africa—that is, the compelling counterfactual, "what might have been"—dissipates, along with the novelty of his captivating terrors. All of this sets the stage for Equiano's relative happiness, which might appear practically plausible given his narrative—let's call it objective—and also subjective and felt, as Equiano himself attests. At the same time Equiano's rather insipid attestation about how he felt might place him on what psychologists call the "hedonic treadmill," which explains why improving circumstances have diminishing returns (the classic examples from the 1970s are tied to the surprising unhappiness of lottery winners and the paraplegic's "surprising" lack of unhappiness[16]). "In contrast to pleasures," summarizes Kahneman, "comforts ultimately produced no significant hedonic experience at all" (13). Or in lay terms we adapt over time, which means that becoming a paraplegic must be distinguished, qualifies Kahneman, from *being* a paraplegic, which is focused around normalcy of a different sort (16). Does hedonic psychology help us understand Equiano's adaptation to the life of a slave after some initial horror?[17]

No doubt Kahneman works out an important awareness, including his observation that counterfactual emotions, such as regret and frustration, guilt and envy, as well as relative happiness, calibrate to what might have been, practically speaking—but also in the sense that speaks directly to the humanities where possibility has greater range than practicality per se. Is Equiano's pleasure in becoming almost an Englishman and his frustration in becoming *almost* an Englishman practical? Says who? Says Equiano in a deeply rhetorical situation writing for Englishmen and women, let us remember. And whence this scale whereby relative Englishness calibrates directly to relative happiness? If that scale exists it is only in the utilitarian imagination that was certainly relevant at the time—like the first edition of Equiano's *Interesting Narrative*, Jeremy Bentham's *An Introduction to the Principle of Morals and Legislations* was first published in 1789 (see chapter 4)—and serves now as an important reference point for Kahneman and his colleagues. In fact Kahneman et al. rely heavily upon Bentham as they resurrect his understanding of experience utility against modern economic theory that considers it unmeasurable.[18] The entire project of hedonic psychology—and much behavioral economics for which Kahneman won the Nobel Prize—might be described as a Bentham rescue effort that would make experience utility measurable but on the newly sophisticated terms of relative not absolute measurement, and thereby render experience a legitimate object of scientific inquiry (xi). So we have a formula that sketches emotion studies across the disciplines: experimental sciences are now centrally interested in the predictably irrational,[19] social sciences in the practically irrational, and humanities in the possibly irrational—that is, how emotions figure the world around limit-cases that qualify our common sense. It turns out that some of the best work in emotion studies worries about how these sometimes separate academic domains impinge upon each other, paying special attention to everyday discomforts that are familiar but don't conform neatly with our academics as they have developed postwar.[20] Now, an example.

Reviewing the difficult question of whether his Good/Bad bipolarity makes sense as a measure of happiness, Kahneman in chapter 1 of his foundational textbook *Well-Being* worries about how to discuss relative pleasures in the same breath with pleasures of any other kind. Can we measure reduction in distress or pain (Bad) on the same scale with pleasures and happiness (Good)? "There is little hope of resolving this problem by introspection," Kahneman admits, "but good reason to believe that the relation between diminishing pain and pleasure will eventually be clarified by studies of relevant brain activity" (9).[21] Coeditor Norbert Schwarz (with Fritz Strack) worries in similar fashion about how the results obtained in different well-being studies offer limited comparability, with cognitive science offering some last-minute hope. Their list of complicating factors is long and wonderfully tied to research studies that admittedly don't add up. They also provide a veritable taxonomy of social experiment rhetoric. Some of the key factors complicating reports of subjective well-being include the following: question-order effects (63), conversational norms (e.g., speakers are expected to provide information that is new to the recipient, 64), irregular chunking of one's life into episodes that color a particular response (65), availability of counterfactuals (67), direction of comparison (68), social improvement expectations (e.g., things will get better, 71), comparison with others who are supposed to share a similar fate (72), attention manipulation in the context of survey or interview (74), and finally, reporting conventions that entail self-censoring (77). "Nearly any aspect of one's life can be used in constructing representations of one's 'life now' or a relevant standard, resulting in many counterintuitive findings" (69), summarize Schwarz and Strack as if they were reading Equiano on his happy slave situation. Subjective well-being reports are not a direct function of one's "objective conditions of life and the hedonic value of one's experiences" (70). Like other social judgments, they are best considered "constructions in response to particular questions posed in a particular time" (79). Indeed. So what should the hedonic psychologists do about this methodological challenge? Like Kahneman, Schwartz and Strack punt as their chapter ends: "If we want to avoid misinterpretations of method effects as substantive effects in this as well as other areas of psychological and social research, we need to learn more about the cognitive processes that underlie the reports that our respondents provide" (80). But instead of deferring to cognitive science as a magic bullet of some unknown futurity, why not further legitimate a methodology around these profound rhetorical insights they offer right now? No amount of cognitive science will ever adequately explain Equiano's happiness, or anything like it. However the hedonic psychologist's deep sensitivity to context, variables—many of them rhetorical—can help in ways heretofore lost on the humanities as well.

As my chapter 4 discussion of Kahneman on mixed feelings will detail, even Kahneman himself shows limited faith in such cognitive science-fiction while he develops the basic outline for a better scientific model that also rejects a common metric of Good/Bad value—a model that does allow for the comparison of values across contexts. For Kahneman, we will find, mixed feelings are not calculable intermediates but rather ambivalencies that reference multiple points subject to what he calls, depending upon the example, negotiation or "persuasion." But for the time being it might be enough to ask the question simply as Kahneman disarms his skeptics: "How happy was Helen in March?" Now tweaked for our purposes: "Was Equiano *really* happy?" That question is not easy to answer and just one of the complications is a matter of time as it takes shape rhetorically. Equiano says he was happy. Or rather he writes that he was happy . . . wrote that he was happy. Or rather he wrote that he considered himself happy. Or rather he wrote that he considered himself happily situated thus implying that in fact he should have been happy given his situation but was not "really happy" whatever that might mean in this case. Finally, at this point in the *Narrative* Equiano might have been situated as a slave merely "by chance"—and not with any particular good feeling or fortune—which in the English eighteenth century served as a synonym now largely forgotten.[22]

Does all of this retrospection and revision and mixed feeling make Equiano the worst kind of unreliable reporter, exactly the kind of reporter scientific psychologists are carefully sidestepping as they develop techniques for "ecological momentary assessment" (e.g., Smartphone survey app + wearable biosensors) designed to eliminate the retrospective distortion of data?[23] If we could hook up Equiano would we want to, and what kind of data would that give us compared to the data that are lost by so doing? In this case I would argue, Equiano's retrospection and revision actually give us the phenomenon in the first place and it would virtually disappear otherwise. An interest in the ecology of emotion is indeed shared by scientists and folks in the humanities, but in Equiano's case the *relevant* ecology is a rhetorical dimension utterly lost on any methodology tied to momentary assessment; in fact, from the perspective of humanities and some social studies research "ecological momentary assessment" is an oxymoron. Whatever the discipline-specific definition, human ecologies must account for temporal factors that exceed the momentary. Then in positive terms: Equiano's case demands rhetorical analysis just as the scientist of psychology Daniel Kahneman implies in his work on well-being *reports*—which are anchored to some specific life domain query—and as he demonstrates explicitly in work on mixed feelings, which accurately describe Equiano's situation at that point in the narrative.

Likewise theorists of situated cognition have typically come up with rhetorical solutions only as last resort, while for those who are more empirically minded, the unity of experience has greater appeal because it holds out the prospect of a common metric. Returning to our foundational excerpt from Merleau-Ponty, they can emphasize the well-being that appears before illness: the intentional arc projects roundabout as our past and our future, bringing about the unity of our senses, intelligence, our sensibility and motility, our "being situated." And in this unity we can establish a common metric for measuring different sorts of beings—no rhetoric needed. Then Alva Noë, in *Out of Our Heads*, emphasizes the complex relationship between the human organism that matures day after day, and the life challenges that disrupt any easy progression and might instigate a kind of retrogression like, we might observe, Equiano's first baptism through the watery middle passage. Perhaps rhetoric has a central place after all.

In *Varieties of Presence*, Alva Noë makes the argument for a rhetoric of experience explicit. Starting with the example of traditional art like song or a painting, Noë explains how mere perceptual exposure is not yet aesthetic experience. Only "through looking, handling, describing, conversing, noticing, comparing, keeping track, we achieve contact with the work/world" (125).[24] But this kind of contact with the world is not neutral; following Kant it falls in the domain of "ought": our response reflects our sense of how one ought to respond to a work of art, for instance. Hence rhetoric as persuasion: "aesthetic experience happens only where there is the possibility of substantive disagreement, and so also the need for justification, explanation and persuasion" (126). Is such persuasive rhetoric relevant only to traditional art forms per se? No—and this is Noë's bold move: he is really working on perceptual experience "*tout court*" (128), with art recapitulating the basic fact about perceptual consciousness and serving as a model or "guide to our basic situation." "Perception is not a matter of sensation; it is never a matter of mere feeling," Noë summarizes. Instead, perceiving is "an activity of securing access to the world by cultivating the right critical stance," or even more direct: human experience has a "rhetorical structure" (128). How do we miss this, according to Noë? "The big mistake," he explains, "is the overlooking of the aesthetic, or critical, character and context of all experience. There is *no such thing* as how things looked independently of this larger context of thought, feeling and interest [classical rhetoric would similarly list the goals of rhetoric: *docere, movere, delectare*]. This is plain and obvious when we think of the experience of art. It is no less true in daily life" (129). So concludes Noë, when something like the Müller-Lyer illusion (lines look different in length even when I know they are the same) is supposed to show that perceptual experience cannot be logically structured in the way that the corresponding belief is structured, we are making a "rhetorical context" mistake.

Noë's interesting narrative takes shape in a philosophical form that could offer Equiano's interesting narrative as a proof text. I will now quote at some length a passage from the earlier work *Out of Our Heads*, because here Noë exemplifies sophisticated situated cognition theory as it appears in a strain of mainstream philosophy that references Merleau-Ponty, as well as J. J. Gibson and Martin Heidegger. Noë thus reinforces the core of my book because his analysis takes for granted some basic comforts as he works, eventually, toward rhetoric. Noë (51 in *italics* below) writes about emergent consciousness and I'll respond with Equiano line-by-line:

> [Noë] *Maturation is not so much a process of self-individuation and detachment as it is one of growing comfortably into one's environmental situation.*

Equiano considers himself happily situated; a contemporaneous account says the Igbo require the gentlest and mildest treatment to "reconcile them to their situation" (243n60). But Equiano has a rejoinder: "Might [African inferiority] not naturally be ascribed to their situation? Does not slavery itself depress the mind?" (45). It is as if this passage were plucked straight out of situated cognition theory à la Noë: consciousness is not something that happens in us; it is something we do in an environment that might be friendly, or I would add following Equiano, might be hostile. The situation where we are "locked in" need not be something innocuous like a dance, to cite a key example from Noë, but rather something terrible like slavery where Equiano is sometimes locked in, or locked up, or locked down, literally.[25]

> [Noë] *We grow apart, but we attach to the world without. We integrate.*

Equiano is almost an Englishman.

> [Noë] *In learning to walk or mastering language, in developing friendships, in acquiring an occupation, in learning to navigate and use technology, we root ourselves in the practical environment.*

Equiano "attaches" to his master, gets his sea legs, learns English, acquires an occupation, learns to navigate a ship and use technology as he roots himself in the practical environment.

> [Noë] *This is one reason, certainly, why radical changes to one's environment, especially occurring later in one's life—for example, in the course of migrating from one country to another . . . are enormous, maybe even devastating personal challenges. The loss of a feature of the environment with which one's daily activities are intimately interwoven is the loss of a part of oneself.*

Before considering himself happily situated, Equiano—relatively late in his life though still chronologically a child between 8 and 11—experiences a radical change to his environment that includes migrating from one country to another.

> [Noë] *Another effect of this kind of disruption is that time, in an interesting, felt way, slows down.*

Equiano's happy situation can appear only after some duration that is marked on a calendar but under the general heading of *a long time*—"between three and four years"—in contrast to the moment-by-moment experience of a terrible rebirth that carries the language of complete disorientation, and "astonishment," when time practically stands still. OED etymology gives us the Latin *attonāre*: to strike with a thunderbolt, stun, stupefy. Obsolete meanings include loss of sense, or insensibility.

According to Noë, and Equiano I add, things take time—and especially for Noë when disruption slows down time in an interesting, felt way. Brain "rewiring" takes time, which is one of the reasons why Noë will look elsewhere to explain a phenomenon like sensory substitution (*Out of Our Heads*, 58). Learning a language takes time while simultaneously it gives us time concepts like dating, counting, and sequencing, which situate us materially (87). Our ability to lock onto and keep track of the world takes time in the form of practical skills (141). What are the feeling structures we find in Equiano's situation? As we have already seen, gratitude appears untenable, just as terror of every kind becomes "strange."

Attachments / Feeling

Time must be taken before Equiano considers himself happily situated. But as we have already seen, time is measured in terms of attachment/feeling. Note in this now familiar sentence the necessary relationship between time ("always") and attachment/feeling: "my master treated me always extremely well; and my attachment and gratitude to him were very great." Noë, for his part, talks about mothers more than masters, but of course the two can blend into one another especially when the topic approaches birth, or rebirth in the case of Equiano. "In a very real sense," explains Noë in the context of finding comfort, "the baby-caretaker 'dyad' is a unity from which the child only gradually emerges as an individual. We can speak of attachment here . . . there is for us no such thing as complete detachment from the community of others and from the larger environmental structures in situations—lights, sounds, odors, ground, the air, technology—up against which we first become ourselves" (*Out of Our Heads*, 51). For better and for worse, according to Noë, we are always already attached to others by the time we can conceive of our ourselves, and likewise for Noë, we are always already attached to larger environmental structures in situations that, I would add, may or may not serve our interests. For Noë as he invokes Merleau-Ponty, attachment is certainly a function of interest, but interest is understood in terms of our everyday and practical involvements: "This is the hand—my hand—whose movements I see when I look. Part of what makes it my hand is that I see it grasping the cup" (74). In other words, for Noë, my hand is "attached" to me not merely because nerve tissue connects it to the brain; the famous "rubber-hand" demonstrates otherwise as the subject can under certain circumstances feel where she isn't attached physically. Instead my hand is attached to me by way of coordination and "common fate." He cites Merleau-Ponty just at this point: "The body is the vehicle of being in the world, and having a body is, for a living creature, to be intervolved in a definite environment, to identify oneself with certain projects and be continually committed to them" (75). Feeling and time.

No doubt feeling/attachment for Equiano speaks in this way to his practical interest as he coordinates on board with his master, and as he shares a common fate that could sink both of them if fate would have it, or could let them live to see another day, with all that entails for these creatures who must make good use of the larger environmental structures in their shared situation: lights, sounds, odors, ground, the air, technology, and so on lists Noë. But would it not be wrong if the interest of the slave is finally identified with the interest of his master? Wouldn't this kind of identification undermine the very distinction between master and slave, which depends upon interests that are vested differently? A fine formulation from Carretta maps Equiano's position with respect to the slave master, as opposed to the overseer who does not share the owners' "vested interest in the well-being of the slaves."[26] Then Carretta gives us Equiano precarious as a freeman still living in an environment where race-based slavery predominates, and where freedom can be rescinded at the drop of a hat because Equiano finds "no one with a vested interest in his well-being" (Carretta biography, 122). Carretta notes the terrible irony that, in some ways, Equiano is more secure as a slave. Security and attachment, in other words, must be understood in terms of vested interests that might work at cross purposes with the feelings involved: for example, Equiano might feel attached to the very set of relationships that subject him.[27]

In fact this question of being was not lost on Merleau-Ponty himself, though it typically is lost by pragmatic philosophers who take him up like Hubert Dreyfus.[28] Merleau-Ponty, in a crucial turn toward the "freedom" that occupies his last chapter, ventures thus: "It may even happen when I am in danger that my human situation abolishes my biological one, that my body lends itself without reserve to action" (84). And emphasizing this Hegelian strand in Merleau-Ponty—the very strand that is lost by pragmatic philosophers—we might consider the slave perpetually "endangered" as he or she is subject to the master within the environment where even a free person is constantly threatened by the institutions and practices of race-based slavery. As Equiano testifies, in this environment the free black can be treated unfairly without much recourse, kidnapped, and resold or sold for the first time into slavery. In this environment interests are vested differently, and life remains unbearable even when a piece of paper attests to equality before the law. Otherwise put, black freedom in this legal sense dissipates into the white world—a world where, to put it mildly, some are more comfortable than others.[29]

Merleau-Ponty, like Hegel, makes the human situation—not biology—fundamental, which is why literature (like Equiano's narrative) can be a good model of consciousness and can provide a rich environment for analyzing experience. "Consciousness takes root in being and time by taking up a situation," explains Merleau-Ponty, and his best example comes from Proust. "Proust shows how Swann's love for Odette *causes* the jealousy which, in turn, *modifies* his love, since Swann, always anxious to win her from any possible rival, has no time to really look at Odette" (425). In this case, "looking" for love takes shape in a manner substantially unavailable to either the cognitive or the social sciences; Proust's novel shows how a particular form of consciousness takes root in being and time by taking up a situation detailed most appropriately in the long form of a literary fiction. Below I will note how Merleau-Ponty qualifies the priority of our human situation over the biological: it is episodic. But first I'll spend some time with Merleau-Ponty's Hegel, who first showed how a human situation like Equiano's might require work in the humanities including the analysis of literature both fiction and "non-" fiction.

Remember for Hegel, in that famous section from the *Phenomenology of Spirit*, master and slave (lord and bondsman) emerge historically as such when a social relationship becomes essentially unbalanced, or to be more precise, when mutual acknowledgment—*Anerkennung*—is foreclosed not just ideologically, but materially.[30] Let us recall Hegel directly. In ¶186 self-consciousness first appears in its immediacy and is not "risen above the bare level of life" (a bit like Dreyfus or Noë on self-consciousness[31]); it is deeply absorbed in the business of living and lacks perspective, as Findlay glosses the passage, because it lacks recognition. Perspective is first achieved, according to Hegel ¶187, as the negation of all mere objectivity, and particularly the negation whereby I am not a thing and I *have* things for myself, by way of objectifying and even destroying others. But to destroy everything else would be to destroy myself as well, since *I am* only by way of acknowledgment, and thus the other must be preserved in some form (¶188). Precisely at this point, for Hegel, the bondsman and the lord appear as such, at the same time that their identity, their original sameness, becomes obscure: "there is posited a pure self-consciousness, and a consciousness which is not purely for itself, but for another, i.e., as an existent consciousness, consciousness in the form and shape of thinghood" (¶189). But crucially, as critical theorists have long read the *Phenomenology of Spirit*, this new master-slave relationship is essentially unequal (¶191), and has within it not only the seeds of a material world but also the seeds of its own undoing. No balance, no pragmatic "equilibrium" or ecological homeostasis here, even after some time. Findlay provides some helpful language as he glosses ¶191: "We thus achieve an essentially unbalanced relationship in which the bondsman altogether gives up his being-for-self in favour of the lord. The lord uses him as an instrument to master the thing for his own (the lord's) purposes, and not for the bondsman's, and the bondsman acquiesces in the situation, and becomes in fact part and parcel of the total objective situation." But because the lord's self-consciousness is now materially bonded insofar as its relationship with the objective situation is mediated by another, it turns out to be the reverse of what it wants—that is, absolute independence. Thus Hegel concludes at this moment in his dialectic ¶193: "The truth of the independent consciousness is accordingly the consciousness of the bondsman."

Noteworthy for our purposes is the essential inequality of this relationship that does not tend toward homeostasis, nor is it a matter of human rights, where human beings initially share everything that counts, and thus renewed balance or equilibrium has a minimum threshold, like a living wage or equality before the law. Instead, for Hegel, the bondsman becomes "part and parcel of the total objective situation." The bondsman does not share with the lord what counts, but instead becomes part of the world in a new way, not comfortably or "deeply absorbed in the business of living," but rather degraded and distorted as part of the objective situation that includes the things made and the worldly stuff encountered as the bondsman goes about his or her business.[32] Both objectively as a thing in the world, and subjectively as consciousness, the world itself appears to the bondsman muted, just as the bondsman appears to the world. This Hegelian mutation could not be righted simply with a living wage or equality before the law, and we see how Equiano grapples with this impossibility. Things don't "flow" or tend toward homeostasis. They cannot resolve through the detached philosophies of stoicism and skepticism in ¶198–¶206, and they cannot be absorbed as the "unhappy consciousness" devoted to "the discordant clang of ringing bells, or a cloud of warm incense" through which the religious beyond is merely felt in ¶217. Nor can the objective world become a sensuous certainty simply looked at, experienced in itself, without having arrived in some form.[33] Instead the Hegelian mutation lives on around the vitality of negation that is not complete. Remember for Hegel complete destruction of the other would be complete destruction of the self because *I am* only by way of acknowledgment, and moreover it is the bondsman who lives most immediately through acknowledgment, though degraded. This is why in the bondsman Hegel finds the (muted) truth of independent self-consciousness.

This is also why Merleau-Ponty, unlike Noë and his more pragmatic colleagues, retains some critical perspective from which normal involvements might appear "queer," if we, like Sara Ahmed, want to make something of Smith's translation from the French.[34] Normal involvements find one part and parcel of the total, objective situation; the retinal image gives us a world right side up, for instance. But all of a sudden the objectivity of that situation can seem queer—in this example when the retinal image inversion experiment ends—or fall apart, "go limp" at the level of my body. But the objective situation also goes limp as illness of the *social* body comes to a head; that is the point Merleau-Ponty makes in an extraordinary, and extraordinarily long three-page footnote at the end of his chapter on "The Body in Its Sexual Being." "As in the case of the individual life, sickness subjects a man to the vital rhythm of his body, so in a revolutionary situation such as a general strike, factors governing production come clearly to light, and are specifically seen as decisive" (172n1). For Merleau-Ponty, insofar as he is a Hegelian in the Marxist line of interpretation,[35] the intentional arc that projects roundabout any particular being, or rather which "results in our being situated," never comes to completion. Hence, being itself can never be self-satisfied no matter how "happily situated" being might appear at any given moment, and this Equiano understood full well, as his narrative demonstrates. In the remarkable passage opening this chapter, Equiano appears much like Hegel avant la lettre, with his happy situation appearing on the page as a set of attachments to the "absolute" of a slave master—service inured, self-considered, well treated, grateful, and almost English—just as the palpable irony of that attachment redounds. Situation "unhappy" indeed. Or from the Hegelian perspective that seems unavoidable in this case, Equiano's interesting narrative allows us to grasp how a happy situation is qualified.

How does Equiano's happy situation measure up to our science of happiness as it is developed by pioneers in the field Daniel Kahneman et al.? Can we make any sense of a scientific conclusion that says somewhere between 10 and 15% variance in levels of happiness is due to demographic factors like ethnicity?[36] And how can we move from an aggregate like this to someone in particular like Equiano, whose happiness variance says a lot but not most obviously "10–15%"? Instead we find that Equiano's happy situation is best understood rhetorically, which distinguishes it from either subjective happiness or "Objective Happiness," as the title of our key Kahneman textbook introduction reads.

"Rhetoric" in this case means a couple of things. One is *effective*, which is to say Equiano's happy situation appears in the narrative consequentially; it does not describe first and foremost. Its design to do something is controlled only partially by authorial intention and discernible by the reader only at the risk of committing the intentional fallacy.[37] Instead most obviously, the passage asks for some kind of a response from the reader, but a response irreducible to simple identification. No doubt such identification is one possibility suggested by reading practices of the time. In his *Rambler* 60 (13 October 1750) Samuel Johnson sketches the proper subject for such narrative writing: "Our passions are therefore more strongly moved, in proportion as we can more readily adopt the pains or pleasure proposed to our minds, by recognising them as once our own, or considering them as naturally incident to our state of life. It is not easy for the most artful writer to give us an interest in happiness or misery, which we think ourselves never likely to feel. . . ."[38] So this theory of identification, implied by Carretta's footnote, would have it that Equiano's narrative of happiness and misery calibrates to the intended reader whose sympathetic passions could produce actions like supporting the abolitionist cause. Questions about practical identification à la Johnson—the similarly classed naturally share incidents—are begged, which leaves only a very different kind of identification before the eyes of God. Even such simple identification, therefore, doesn't strictly require sincere emotion reported transparently; required is only that the reader adopts some modest "proposal." But at the same time identification is not so simple. A reader moved to abolition by Equiano's narrative need not feel happy or miserable with Equiano, whom we are not sure feels one way or another anyway: he qualifies, ironizes, looks back and forth artfully. Moreover a reader's response need not identify; Aristotle understood that tragedy can be a source of pleasure for instance. The design itself carries rhetorical force by inhabiting conventions. At the same time the author Equiano figures the convention ironically, which makes sense only in relation to his life story, which is to say you have to read the story or at least know the setup for the irony to appear. The emotion is rhetorical in that it requires, to appear as such, a reader or audience broadly conceived.

Second, "rhetorical" in this case means that the emotion—*pathos* in the classical lexicon—is not first and foremost something Equiano feels, for example, happy, nor is it something observed like a smile or a positive disposition; instead, it posits a social relationship. Notice that Equiano does not feel happy but instead finds himself, ironically, in a happy situation, which posits an attitude toward: the scene of subjection[39] (~terror), the master-slave relationship (~gratitude), and a type of social identity (~quite English). In other words, this happy situation would be lost on anyone who isn't paying attention; it is rhetorical—and it cannot function primarily as an assessable state-description. Even Kahneman's "dense record of the quality of experience at each point-instant utility" could not give us an assessment of Equiano's objective happiness over a period of time (3). And it would be an understatement to say that Equiano's positive emotions do not unequivocally "signify that life is going well, the person's goals are being met, and resources are adequate," as Ed Diener and his colleagues might surmise on their way to a theory of the happy human organism.[40] More promising is Kahneman's work on how well-being reports are *anchored*, which is precisely the sort of rhetorical analysis invited by Equiano's narrative, as I'll demonstrate (see chapter 4).

So far we've looked at how time cannot add up for Equiano, and how attachment comes at a price—one might call these Hegelian or "negative" phenomena. But what about the world as given and the things in it? Isn't this how, as pragmatic philosophers might argue, we must find our comfort zone even in the most trying of times? Things normalize, don't they, because an unbearable situation is unsustainable? Next we will take a look at how, for Equiano, the world and all things do indeed "bear up." And once again my approach will be a kind of critical phenomenology, where in this case I pay attention to the ways in which a world and all things can be given (to a slave).

Bearing Up

What appears given? For the philosopher Alva Noë, what is given depends upon the relationships we bear, including human and nonhuman: "We bear a relation to the house: the house itself, the back no less than the front (or the front, no less than the back), is within reach" (*Out of Our Heads*, 83).[41] Similarly Ahmed starts with philosophical discourse and she pays careful attention to what is given virtually: "picture this . . . ," "For instance I see that . . ." (25); objects gather along certain lines not others. Then following Merleau-Ponty, Ahmed moves in exemplary fashion from neutered philosophical discourse to sexuality where things appear different. Now the world of heterosexuality appears given, the object of desire appears this way not that (87). Moreover in a world where heterosexuality is normative, Ahmed qualifies, some objects not others appear to be the appropriate object of desire, which also means that we will have experiences that register precisely by way of their orientation, which may reference the norm even when failing to orient properly. In the tradition of critical theory running from Louis Althusser through Judith Butler, Ahmed can also put this in terms of sense perception: our orientation depends upon taking certain points of view as given. Certain things and not others become available and come into view—the subject turns, or is turned, one way and not another (14, 165). Now you might guess where I am going with this, as I return to Equiano.

A methodological point: This is one way we can read narrative literature, and Equiano's interesting narrative in particular. We can ask: What appears given to Equiano? What objects appear in the narrative and what does their appearance say about Equiano's orientation in a world turned this way not that? How are givens revealed by tracking what Equiano senses? As we see in this chapter by way of Equiano, narrative literature does an especially good job reconstituting and hence revealing for analytic purposes the complex relationships between the objective world and experience, as this relationship takes shape in character. In a formula, books are orientation devices—they do something—and some books are largely about just this orientation process. So, for example, Equiano's *Interesting Narrative* is abolitionist, and the way it is abolitionist is by offering the orientation that makes abolition necessary. If you read the story, then abolition makes sense. The strong argument is that a legal event—abolition—can appear as such only when oriented, and orientation in this case requires not just some version of a mariner's quadrant, but a narrative. Let us see how this works in detail, as Equiano gets his bearing.

The world bears down upon Equiano after he is "deceived by a promise of being delivered" from slavery and instead finds himself sold to one Captain Doran sailing for the dreaded West Indies. "In the first expressions of my grief I reproached my fate, and wished I had never been born. I was ready to curse the tide that bore us, the gale that wafted my prison, and even the ship that conducted us; and I called on death to relieve me from the horrors I felt and dreaded" (97). In this pair of beautifully adjacent sentences, Equiano bears witness to his sense of being put upon, most dramatically. What's going on here stylistically, and does it matter? I'll argue quite a bit and yes; the *Oxford English Dictionary* sustains this linguistic network that Equiano's narrative brings to bear in almost every sense, we will see, including the alliterative sense born/bore that might seem merely stylistic. So now a brief OED excursus, edited for our purposes.

> OED Bear (v.). A very old and widespread root across English and related languages. Main senses: I. to carry; II. to sustain; III. to thrust, press; IV. to bring forth. 2b. To bear in mind. 2c. To bear witness. 3. To bear the cross [of Jesus]. 6a. To bear arms against. 7b. To bear a name or 7c. an office. 9. To entertain a feeling, e.g. 1726 Swift *Gulliver* II. iii. ii. 27 "The Contempt they bear to practical Geometry." 17. To bear with, be patient. 21. To bear up. To keep up one's courage or spirits; to maintain one's ground, e.g. 1796 E. Burke *Lett. Peace Regic. France* in *Wks.* (1842) II. 291 "Bearing up against those vicissitudes of fortune." 27. To bear down. 32. To bear upon or touch upon. 33. Bring to bear. 37. In nautical phraseology, to sail in a certain direction e.g. bear away, bear down, bear off, bear up. Bearing (n.) 1. The action of carrying or conveying. 2. The carrying of oneself, deportment, e.g. 1600 Shakespeare *Much Ado about Nothing* ii. i. 150 "That is Claudio, I knowe him by his bearing." 5. Sustaining, supporting, e.g. 1733 Pope *Ess. Man* i. 23 "Of this vast Frame the Bearings, and the Ties." 9. Tendency to exert influence, e.g. 1785 E. Burke *Speech Nabob Arcot's Debts* in *Wks.* IV. 201 "Having had . . . a just sense of their true bearings and relations." 13. To take one's bearings: to determine one's position with regard to surrounding objects; also *fig. 17. Child-bearing.*

So Equiano's tight alliteration in fact exploits morphology; born/bore initially gives us two interlocking dynamics. First it gives us a particular way of (not) being in the world practically, as one bears a name. "While I was on board this ship, my captain and master named me *Gustavus Vassa.* I at that time began to understand him a little, and refused to be called so, and told him as well as I could that I would be called Jacob; but he said I should not, and still called me Gustavus; and when I refused to answer to my new name, which at first I did, it gained me many a cuff; so at length I submitted, and was obliged to bear the present name, by which I have been known ever since" (64). From the equivocation in the book title—*Equiano or Gustavus Vassa*—his name bears practical witness to the profound problem of knowing who he is. Not in some self-help sense that grounds us, or grounds Equiano and bolsters his self-confidence as he waits for a chance to realize his dreams: a "self-made man" as we read in the title of Carretta's biography. Equiano does not know who he is precisely insofar as he is excluded from the systems of knowledge that make one's identity intelligible. Not free, not equal, not quite an Englishman, indeed not human in the ways that count. And not ever, as Afro-Pessimism has recently argued in the face of an affirmation that would give black people a piece of the action.[42]

I like to imagine Equiano recognizes some bitter humor in these circumstances when he includes verbatim in his narrative not one but two "certificates of behavior" that bear his name (again our keyword). The first certificate Equiano requests from Robert King: "I then requested he would be kind enough to give me a certificate of my behaviour while in his service, which he very readily complied with, and gave me the following." The certificate reads:

> **Montserrat, January 26, 1767.**
> The bearer hereof, Gustavus Vassa, was my slave for upwards of three years, during which he has always behaved himself well, and discharged his duty with honesty and assiduity.
>
> **ROBERT KING.**
> **To all whom this may concern.**

And once again Equiano appears truly grateful: "Having obtained this, I parted from my kind master, after many sincere professions of gratitude and regard, and prepared for my departure for London" (163–164). In this case the system of knowledge that makes Equiano's identity intelligible—the certificate of behavior—bears upon him as the most flimsy sort of attachment, a written voucher referencing a certain Gustavus Vassa, who supposedly behaved himself well as a slave, whatever that means, for whomever it may concern. At the same time this certificate is worth its weight in gold and for it, Equiano expresses the deepest gratitude. What kind of bearing upon the world is this?

The natality-sustainability dynamic is also practical with respect to Equiano's familiar surroundings. As the tide bore Equiano in his ship, which is to say "prison," he is wafted about as he loses his bearing with respect to everything that gave him hope: the "wished-for" land he loses sight of, British law banning slavery on the military ship where he had just served, the law prohibiting a sale like the one he has just suffered, the financial equity he thought earned toward his self-purchase (93), and most importantly his master Pascal's promise that Equiano might be freed. We can gloss this broken promise with Ahmed, who helps us understand how at this point Equiano is undone not just by a promise gone awry, but by his very exclusion from the social world that might give his servitude some kind of meaning however debased. Following directions, explains Ahmed, normally takes the form of commitment, a social investment that promises return (17). If we fall into line as we are pressed into some kind of social service, our commitment will be rewarded if only by way of some recognition that we are living in a certain way. But what if we are pressed into service outside the domain of consent or even beyond the social contract per se? Here we must amend Ahmed. For in Equiano's case, the service into which he is pressed bears false witness to the promise that he would be freed after fulfilling some kind of social contract, whether that means becoming almost an Englishman, buying his freedom, or even, via baptism, cleansing his soul before the eyes of God and thus joining the Christian fellowship where everyone is equal. No wonder Equiano was ready to curse the tide that bore him, even to curse God, as Equiano wished he were never born.

And this gets us to a second, crucial dynamic at work in this passage. Equiano takes his bearing with respect to the practical world where he is sadly disappointed to the point where he curses his own birth. But then again this passage is given to the reader in retrospect and with qualifications: "in the first expressions of my grief [but not thereafter] I reproached my fate . . . I was ready to curse [but did not]." So at the same time that Equiano bears witness to his life-threatening misery as it is attached to the practical world that pressed down upon him, he qualifies that misery as something he could ultimately bear, as in "carry" (OED main sense I). Clearly Equiano sustains himself through this episode (OED main sense II) since we are reading his retrospection. Ultimately he is patient (OED I.17). One might even say his suffering at this point shows Equiano bearing the cross for Jesus in a manner he could barely recognize at the time. So we should now take another moment to qualify Ahmed. No doubt Equiano's practical world provides orientation, or in this case disorientation. Ahmed might say body and object "point" toward each other along particular lines; Equiano is oriented by how he takes up time and space, which in this case undoes him (Ahmed 5). He can't bear it. However, at the same time, he is telling this story about the cross he could bear after all, a story where he regrets the practical orientation that almost undid him. And that's the point of this story: sometimes a perfectly good practical orientation can be deadly, while life itself can hinge upon the impractical.

Finally we might wonder, What exactly is the phenomenology of Equiano's born-again Christianity? And how might it square with our practical imperatives as they come out of science and philosophy? Equiano, having purchased his freedom and converted, embarks March 1775 on board a ship bound for Cadiz (198–199). The ship strikes a rock and although Equiano cannot swim he feels no dread in his "then situation" while these words come to mind:

> Christ is my pilot wise, my compass is his word;
> My soul each storm defies, while I have such a Lord.
> I trust his faithfulness and power,
> To save me in the trying hour.
> Though rocks and quicksands deep through all my passage lie,
> Yet Christ shall safely keep and guide me with his eye.
> How can I sink with such a prop,
> That bears the world and all things up?[43]

My plea at this point is we have to take seriously such practical limitations, or to put this another way, we have to take seriously such impracticality where Christ (not I) bears the world and all things up. In fact for Equiano, such impracticality is a matter of life and death, as we see him narrate over and over. The fullest version of this life-and-death narrative comes in the set piece about Equiano's spiritual rebirth that I reproduce here with annotation and underscoring of what Merleau-Ponty might consider the intentional arc rehabilitated after going limp ("It is this intentional arc which brings about the unity of the senses, of intelligence, of sensibility and motility").

> In the evening of the same day, as I was reading and meditating on the fourth chapter of the Acts, twelfth verse [Peter's miraculous escape from prison], under the solemn apprehensions of eternity, and reflecting on my past actions, I began to think I had lived a moral life, and that I had a proper ground to believe I had an interest in the divine favour; but still meditating on the subject, not knowing whether salvation was to be had partly for our own good deeds, or solely as the sovereign gift of God; in this deep consternation the Lord was pleased to break in upon my soul with his bright beams of heavenly light; and in an instant as it were, removing the veil [also a metaphor of Du Bois], and letting light into a dark place, I saw clearly with the eye of faith the crucified Saviour bleeding on the cross on mount Calvary: the scriptures became an unsealed book, I saw myself a condemned criminal under the law, which came with its full force to my conscience, and when "the commandment came sin revived, and I died" [Romans 7:9 + 7:25 "I myself in my mind am a slave to God's law"], I saw the Lord Jesus Christ in his humiliation, loaded and bearing my reproach, sin, and shame. I then clearly perceived that by the deeds of the law no flesh living could be justified. I was then convinced that by the first Adam sin came, and by the second Adam (the Lord Jesus Christ) all that are saved must be made alive. It was given me at that time to know what it was to be born again, John 3:5. I saw the eighth chapter to the Romans [8:14 "For those who are led by the Spirit of God are the children of God. The Spirit you received does not make you slaves, so that you live in fear again; rather, the Spirit you received brought about your adoption to sonship"], and the doctrines of God's decrees, verified agreeable to his eternal, everlasting, and

> unchangeable purposes. The word of God was sweet to my taste, yea sweeter than honey and the honeycomb. Christ was revealed to my soul as the chiefest among ten thousand. These heavenly moments were really as life to the dead, and what John calls an earnest of the Spirit. This was indeed unspeakable, and I firmly believe undeniable by many. Now every leading providential circumstance that happened to me, from the day I was taken from my parents to that hour, was then in my view, as if it had but just then occurred. I was sensible of the invisible hand of God, which guided and protected me when in truth I knew it not: still the Lord pursued me although I slighted and disregarded it; this mercy melted me down. When I considered my poor wretched state I wept, seeing what a great debtor I was to sovereign free grace. Now the Ethiopian was willing to be saved by Jesus Christ [Acts 8:26], the sinner's only surety, and also to rely on none other person or thing for salvation. Self was obnoxious, and good works he had none, for it is God that worketh in us both to will and to do. The amazing things of that hour can never be told—it was joy in the Holy Ghost! I felt an astonishing change; the burden of sin, the gaping jaws of hell, and the fears of death, that weighed me down before, now lost their horror; indeed I thought death would now be the best earthly friend I ever had. Such were my grief and joy as I believe are seldom experienced. (189–190, my underline)

Let us remember, first, this is a passage from a late chapter Mary Wollstonecraft famously panned after praising Equiano's first volume with its "blood-turning course." For Wollstonecraft the latter part of the second volume reads flat and she finds "the long account of his religious sentiments and conversion to Methodism . . . rather tiresome."[44]

No doubt Wollstonecraft is right that Equiano's conversion narrative takes conventional form as it might appear in the hands of any number of spiritual autobiographers. It's a set piece in some ways, and reduced to its essence, the allegorical structure of salvation runs the normal course: 1) godly sadness, 2) rebirth, 3) challenge, and finally 4) manifestation of the new creature. How exactly this passage "gives things" to Equiano is striking, however, and it brings home our reading of the happy situation where we first found Equiano.

Most importantly in this scene we witness how Equiano's senses—touch, taste, sight—are "unified" as Merleau-Ponty would have it, beyond the practical world where life is impossible. Equiano dies, he is reborn. But is this initial death a kind of spiritual illness before health is restored, do we witness Equiano going limp and losing his way before finding it once again? Is Equiano happily re-situated as the philosophers and psychologists would have it? The trick here, I think, is that Equiano's recovery and the unity of his senses cannot be borne by himself, since that option is available only to someone who is already in comfortable conversation with the world as such. We have seen how Alva Noë has claimed only "through looking, handling, describing, conversing, noticing, comparing, keeping track, we achieve contact with the work/world" (125). But what is this "rhetorical structure" of human experience when we are not already in comfortable conversation with the world? When the conversation is awkward at best and foreclosed in fact? A litany of the inoperable: Equiano reminds us that he lives in a world where a master might enslave his own progeny and hence the normal rules of blood filiation don't apply (109); where "every principle and all sentiments of virtue" are buried (110); where equality and independence, those "first natural rights of mankind," are violated (111); where there is no common ground or common currency through which one might communicate "one rational being to another" (112); and where black speech can be better described as black infelicity—Equiano reminds us that throughout the West Indies "no black man's testimony is admitted, on any occasion, against any white person whatever," which means that he had no recourse when his "kind master" Robert King shuffles instead of repaying a financial debt that Equiano desperately needs (161–162). Again, what is the rhetorical structure of human experience when the world is not conversable?

The world as fallen for a slave, and Equiano in particular—the escaped prisoner, the slave to God's law, the Ethiopian—is fallen differently than it is for another kind of sinner; hence rebirth differs as well. As a slave, Equiano bears down on a world that is not his own; then Jesus bears everything. Suffice it to say in summary this is a very different kind of phenomenology that loses its bearing outside of the rhetorical situation broadly conceived, including both the no-go zones of common currency and conversation that would (but doesn't) give Equiano a voice in some traditional sense of public sphere representation. Also his loud statement as we have received it—his *Interesting Narrative*—constantly challenges his reader to take the material at face value even when he demonstrates over and over how that value is in some basic ways phantasmatic and impossible to cash. We cannot bank upon the fact that Equiano is really happy, truly grateful, and so on through the narrative, as we are time and again thrown back upon the system that makes such guarantees impossible. In the end Equiano is neither here nor there, neither secured in the practical world where he does his business, nor in the spiritual world where he tends. As such the narrative Equiano gives us is not tiresome in the least, recalling the words of Wollstonecraft, but vivifying as only a narrative of such improbability can be. And no further explanation should be needed how a narrative such as this speaks directly to a science of emotion, without reducing to it.

3

Hostile Environments in Ann Radcliffe's *The Romance of the Forest*

In this chapter we see how environments, technically understood via J. J. Gibson, make people terribly uncomfortable to the point of becoming unhinged. Or even "undone," as rape victim Antonia protests in Matthew Lewis's *The Monk* (1796) just before narrative logic finishes her off: "Is not my ruin completed? Am I not undone, undone for ever?" In fact the cause of death in this instance, and more generally the terms of a life undone, find their genre in the sentimental version of Gothic fiction where the emotional scene is troubled. A narrative logic that Walpole, Monk, and especially Radcliffe pursue vigorously shows how—in contrast to tragedy, which undoes through character—domestic "scenes" and other human environments can pose a real threat. In this chapter I will thus investigate how certain kinds of narrative fiction provide rich material for the study of human environments, especially in terms of the emotional dynamic. In this first section, I detail some neighboring approaches so that the contrast with a more ecological Gibsonian model is clear, followed by a reading of hostile environments as they are defined legally in our contemporary workplace environment, and as they appear in Ann Radcliffe's Gothic fiction. My ongoing argument is first, that it matters very much what model of cognition one uses when studying how human environments appear in literature, and second, that this specialized literary critical issue in fact lies at the crux of a significant encounter between science and humanities.

Limitations of Literary Darwinism and Cognitive Approaches to Literature

Literary criticism is typically a special interest. But nonspecialists have also paid attention to both literary Darwinism and cognitive approaches to literature, where basic methodological issues are worked out with consequences for ongoing research, and even for certain kinds of social and educational policy.[1] Literary Darwinist Brian Boyd, for instance, proposes that an evolutionary view allows for "*informed* social change" (27, emphasis original) as he introduces his substantial account published in 2009 by the Belknap Press of Harvard University, *On the Origin of Stories: Evolution, Cognition, and Fiction.* What would this kind of social change look like? It is difficult to say because unlike Martha Nussbaum, who has parlayed a cognitive approach to literature into controversial social policy proposals,[2] critics like Boyd are typically demure as they let the reader extrapolate what kind of social change would be informed by the human nature they sketch. To that end we are offered some hints however: "Each offspring in our species should prefer to concentrate its parent's investment in itself," explains Boyd, while its parents "should prefer to maximize their genetic transmission to the next generation by offering, on average, equal investment in each of the present and future offspring" (55–56). What kind of informed social change might follow from this rehearsal of the literature on kin selection? Although Boyd doesn't say exactly, he expects us to extrapolate in this case from the principle of kin selection to expressions of human nature, whether that means a social institution like marriage and family law, or an artifact like fictional narrative. "Hence we see why even in the most cooperative of relationships competition is inevitable, and why the powerful emotions engendered by family loyalty and conflict saturate stories from Genesis to *The Sopranos*" (56). This great leap from the theory of kin selection to HBO leaves us to infer by way of analogy that the relevant social change would . . . (try finishing the sentence because I'm at a loss). Another would-be clue turns up in the section on "Fiction's Functions" where Boyd discusses how fiction improves the key mental mode of social cognition along with subsidiary functions arising from that fact. He gives us this curious sequence before breaking off: "Males have more reproductive variance than females . . . therefore have on average a stronger drive than females to earn status . . . and therefore males [despite the likes of Murasaki, Jane Austen, and J. K. Rowling] are overrepresented at both extremes, success or genius, and failure, crime, mental illness, or drug dependency" (195). Again what kind of social change might be informed by this evolutionary view is, perhaps, better left unsaid. However, Boyd doesn't miss the opportunity to tell us what kind of social change this *won't* be: Enlightenment blank slate, romantic à la Rousseau (65), Marxist (387), or "paranoid" in the traditions of Nietzsche and Foucault (66). The claim is that "Evolution offers a much more complex and nuanced view of the social world" (65). But if evocritical and some cognitive approaches to literature are any indication, this promise of a better theory of social change starts on unstable ground before it would have to navigate a rather daunting history of social Darwinisms gone awry. Perhaps that is why for the time being these critics are sticking to the arts—controversial enough—while leaving vague the political program.

J. J. Gibson's affordance theory provides a welcome alternative to certain Evocritical and Cognitive Approaches that are hampered by the operative model of cognition that toggles between the poles of a prehistoric architecture on one hand, and neuro-"plasticity" or flexibility on the other: a polarity that over- or under-shoots the analytic level appropriate to our phenomenal world as evident in literary artifacts. Again Boyd: "A biological view of human nature stresses that individuals are free agents endowed with the flexibility that evolution provides and active strategic choosers rather than passive products of their place and time." Meanwhile culture "occurs only within the social and therefore, again, the biological realm" (25), which means that the force of cultural and critical interpretations of literature are blunted as they run up against the latest biological fact. Despite protests to the contrary which I will revisit below, literary-critical paradigms (like psychoanalysis) that are grounded in the humanities, not biology, get squeezed out of the picture. "As a theory," Blakey Vermeule writes candidly by way of example, "psychoanalysis is undeniably rich. As a story about the mind, however, it is laughable."[3] Another leader in these critical movements on the cognitive side, Lisa Zunshine, worries more about losing her academic neighbors in the English department, so her model is instead additive. In her *Introduction to Cognitive Cultural Studies*,[4] Zunshine insists that a cognitive turn in literary criticism does not mean abandoning the traditional paradigms be it gender studies, feminist criticism, postcolonial theory, poststructuralism, performance theory, psychoanalysis, or cultural studies (4). But though these paradigms might reappear in the critical literature, I will argue that they do so on scientific grounds thereby undermining what is meant to be collaborative work that respects the domain-specific relevance of the various critical paradigms. So when Ellen Spolsky writes in the same Zunshine volume about "Darwin and Derrida," for instance, troping, reinterpretation, and representation are understood not as literary devices with all the work that they can do in the spirit of Paul de Man or Jacques Derrida, but rather as—what else—"adaptations" consistent with a Darwinian theory of survival (306).

Indeed Cognitive Approaches, along with its critical neighbor Literary Darwinism, get much of their popular appeal from claims about human nature, especially as they trigger some of our heated conversations around the special topics of sex and gender, selfishness, and the biological origins of social life. At one end of the spectrum in literary studies we have a movement that wants to explain how cultural phenomena like art and literature should best be understood as the product of biological evolution. Boyd, in a response to Jonathan Kramnick's *Critical Inquiry* (*CI*) polemic "Against Literary Darwinism" (2011), provides the following definition of the movement that he prefers to call evocriticism: "Evolutionary criticism argues that evolution has shaped human minds to be partially reshaped, not least by our species-wide predisposition to culture, to art in general, and to literature in particular. Evocritics show how the fact that human minds owe their structure to evolutionary pressures makes a difference to literature—to features of human minds and behaviors that literature deploys, represents, appeals to, engages, and modifies" (*CI* 394). I will return to this definition, though a response to Literary Darwinism per se is not my objective; I focus, instead, on models of cognition that in some cases reject Literary Darwinism outright.

Though Boyd and many of his colleagues maintain that evolutionary and cognitive approaches are only "different facets of a common process of inquiry" (9), they are in fact not identical, and I will emphasize how they differ significantly at the extremes where my critical work might be situated toward one end of the spectrum with evolutionary approaches at the other. As demonstrated in the definition above, moreover, Literary Darwinism has largely dissociated from the more radical forms of "orthodox" Evolutionary Psychology (EP), defined best by Leda Cosmides / John Tooby, David Buss (*CI* 397), and Steven Pinker (*CI* 406), while identifying with a broader field of evolutionary inquiry into human nature referred to as evolutionary psychology (ep, without caps). The key difference lies, it would seem, in three arenas worth rehearsing in the spirit of serious engagement. First is ep's new commitment to brain and mind plasticity or flexibility (399, 407), which is also a central tenet of recent cognitive approaches. Second is "multilevel selection," which acts above the gene at the level of organisms and groups, allowing an evocritic like Boyd to explain how fictions can enhance social cohesion while he still can appeal to biological explanations (*CI* 399–400). Third is gene-culture coevolution, where a paradigm for Joseph Carroll is cooking (i.e., culture), which freed up for our prehistoric ancestors metabolic resources for brain development—a paradigm that would extend to literature and the other arts (i.e., culture), though he does not let on how the analogy falters on genetic influence. There is no evidence that literature and the other arts have, like cooking has, altered the human genome. Though it takes the form of a causal claim, it is difficult to understand how exactly art "transforms our instinct-driven organism into an organism that makes decisions based on complex mental representations of the world,"[5] and in fact it is difficult to imagine exactly what kind of empirical evidence would make this quite specific causal claim testable. Though perhaps unfair, at this juncture I like to imagine a *New Yorker* cartoon with slime mold sitting in a comfy chair reading Bonner's *The Social Amoebae.*

Hence a note on Literary Darwinism and scientific testability as opposed to some other evaluative standard. Transitioning to Book II, "From Zeus to Seuss: Origins of Stories," Boyd summarizes his naturalistic account of art and fiction where he argued that both are adaptations: "we have evolved to engage in art and in storytelling because of the survival advantages they offer our species" (209). This and related hypotheses "need to be tested empirically against alternative explanations," Boyd continues. "But an evolutionary approach to literature does not depend on them or on any other hypotheses that claim art, literature, or fiction as adaptive. A biocultural approach to literature [P] simply requires that [Q] we take seriously that evolution has powerfully shaped not just our bodies but also our minds and behavior. We can do that whether literature or fiction is an adaptation, byproduct, or some combination of the two" (209–210). Logicians call this fallacy *affirming the consequent* (if P, then Q. Q. Therefore, P): If you are going to accept approach P, then Q is required. Q is required; therefore you must accept approach P. But in fact you may take seriously as I that evolution has powerfully shaped "not just our bodies but also our minds and behavior" without in any way committing oneself to a biocultural approach to literature, which is something else altogether. Wildly swinging at Theory with a capital "T," Boyd stumbles down a similar path in his conclusion: "Unlike current Theory, evocriticism prefers proposals concrete enough to be subjected to potential falsification by evidence" and he offers that aspects of his own proposal for art as adaptation could be falsified, for instance, if there were on average a negative correlation between individual artistic success and status (388). But when it comes to testing the validity of his literary criticism, Boyd admits that the evaluative criteria change: "Claims within an evolutionary approach to literature can hardly be 'proved' by application to literary works. Yet the fertility of such an approach should be tested both on particular works and on literary theory. If evocriticism cannot generate literary theory and criticism richer than those possible through other models, it does not warrant the considerable effort of learning enough about human evolution to apply it to literature" (388–389). Wisely enough, evaluation of evocriticism per se hangs on judgment about relative richness—never mind equivocation around the word "test"—and on these terms I venture the ultimate judgment will not be kind.

Taken together, all three retreats from orthodox EP would nevertheless immunize Literary Darwinism against its critics like Kramnick, at the same time that they would make room for criticism focusing on some of the finer details of culture, politics, and creativity as they appear in literature. I will not highlight the persistent shortcomings of Literary Darwinism even after this turn away from orthodox EP, since those disputes are well-summarized in the exchange with Kramnick, and because my focus here is on operative models of cognition in current literary criticism. Though important, ongoing debates about art as either an evolutionary engine or byproduct, group selection, or the definition of "culture" in gene-culture coevolution, won't preoccupy insofar as the controversies are inherent to the theory and science of evolution. However, when it comes to the operative understanding of cognition, we will see, new ep finds significant overlap with CAL, especially in the appeal to neuroplasticity as a black box supervenient upon genetics. I will argue that cognitive approaches to literature, like ep, frequently preclude robust critical theories, despite efforts by some to honor these literary critical traditions and appease skeptics. I will return to ep for another reason as well, as it offers a competing model of how and why hostile environments are depicted.

For her part, Zunshine in her introduction to the *Oxford Handbook of Cognitive Literary Studies* is adamant about the difference between cognitive literary critics and Literary Darwinists, rejecting consilience as it is embraced by Boyd in particular.[6] However these two critical approaches to literature frequently share methodological problems more trenchant than Zunshine admits. The way Zunshine puts it, a "distinction between cognitive literary critics and Literary Darwinists is worth emphasizing because, while both draw on some of the same research in cognitive science (e.g., evolutionary theory), their views on the role of this research for literary studies are diametrically opposite." Here my point is to reverse this emphasis showing how shared cognitive science research in evolutionary theory can lead to some shared problems with respect to literary criticism. These two movements—CAL and ep—overlap significantly when they try to show how, in the words of Boyd, the fact that "human minds owe their structure to evolutionary pressures makes a difference to literature" (*CI* 394).

Let us grant that human minds owe their structure to evolutionary pressures, and that literature is the product of human minds. Still the specific charge is not met: namely to demonstrate what *difference to literature* such evolutionary pressure makes. To do this, a critical difference in literary interpretation must be identified, and this identification must be best explained by way of an appeal to evolutionary pressure with enough detail to mark this difference or class of differences, as opposed to others. The counterclaim is that we are not talking about the possibility of a scientific demonstration of causal connection, where some kind of archaeological or laboratory data could demonstrate that evolutionary pressure *x* produced the difference to literature *y* (nomadism produced the travel narrative, or what have you). Instead we are talking about what Jerry Fodor would call post hoc "historical narratives" that are more or less plausible, more or less convincing. In his polemic "Against Darwinism," Fodor calls this a pragmatic or "rhetorical conclusion" where the parameters around an explanation depend upon "what is being explained and to whom, and to what end."[7] So at least when it comes to procedures for assessing the claims of Literary Darwinism, Fodor and Boyd might agree despite appearing on opposite sides of the adaptation argument: neither would think Darwinian literary criticism can be tested according to the strictures of scientific method.

For the sake of a focused argument, my contention in this chapter is that the biology-literature connection is least convincing when a favorite target of such criticism—narrative fiction and especially novels—are read as the expression of distant evolutionary pressure, and more convincing when read as proximate human ecologies. But before such focus, some of the more sprawling claims and issues must be brought to bear as they muddy the water unavoidably.

Any approach that initially separates species and environment, or biology and history, or mind and world, before stitching them back together loses sight of the phenomenon at hand. Hence a list of red flags with respect to literary criticism:

- Pleistocene, Hunter-gatherer (remote from the literary artifact)
- evolved mind, or evolved mental architecture (question begging)
- consilience (reductive)
- modularity (deterministic)
- neuron (too small)
- plasticity (black box, too big)

Now let us see in more detail how key examples of such literary criticism tend to proceed. At the heart of CAL we find Ellen Spolsky's position statement in her preface to *The Work of Fiction* (2004) reiterated by Lisa Zunshine in her *Introduction to Cognitive Cultural Studies*, which, despite the title, also focuses on narrative fiction.

> Cognitive literary theory is . . . well positioned to provide insights into a question that has been occluded by the well-deserved successes of the reemergence of historical and multifaceted cultural studies that has proliferated after the New Criticism in the twentieth century. That question is this: how does the evolved architecture that grounds human cognitive processing, especially as it manifests itself in the universality of storytelling and the production of visual

> art, interact with the apparently open-ended set of cultural and historical contexts in which humans find themselves, so as to produce the variety of social constructions that are historically distinctive, yet also often translatable across the boundaries of time and place? It is the job [of cognitive literary theory] to begin to chart the emergence, manifestation, and readability of these only temporarily stable relationships between the humanly universal and the culturally and individually specific, as coded and recorded in cultural artifacts. (2)

Reading quickly one does get the impression of a complex relationship amongst brain-being-world, where "cognitive processing" is the brain understood computationally as information processing, "evolved architecture" is the human being understood as a species characterized by storytelling and artistic adaptations, and "cultural and historical context" is the world in which individuals find themselves, piecing things together in what turns out to be the unique cultural artifact such as Ann Radcliffe's *Mysteries of Udolpho* (briefly discussed by Zunshine, 128), or for that matter the *Communist Manifesto*, or Picasso's "Guernica," or the cave paintings at Altamira. In principle no major field of study would be left out, including serious science, social studies, and the humanities. Or to put this another way, the skeptic appears stymied because cognitive literary studies à la Zunshine denies no academic first principles, and we will see how this hasty inclusiveness works in an example below. The physicist, the chemist, the biologist, the historian, the sociologist, the Marxist, the feminist, the formal critic, and so on, might all be accommodated as one moves up and down levels of analysis from the microphysiological (e.g., Candace Pert's *Molecules of Emotion*, which is cited in the literature, or Alan Richardson's *The Neural Sublime: Cognitive Theories and Romantic Texts*) to the highest art. "We still need digestion in order to enjoy Bach," remarks cognitive neuroscientist and authoritative reference point Antonio Damasio,[8] shorthand for integration that excludes nothing except perhaps for an approach, like Cary Nelson's Cultural Studies, that itself excludes consideration of "the evolved human brain" (a key term used, we often forget, by New Left literary critic and Cultural Studies founder, Raymond Williams).[9] Despite protests to the contrary, a project under this description may appear consilient along the lines of biologist E. O. Wilson, or evolutionary psychologist Steven Pinker, whom Zunshine cites as a fellow traveler through the *Introduction to Cognitive Cultural Studies* collection (311n2).

"Consilience," also discussed in the final chapter of my book, would explain any human phenomenon in natural-scientific terms. Thus, according to Pinker for instance, a novelistic emotion like romantic love in Jane Austen's *Sense and Sensibility* should be read in terms of evolutionary psychology that includes mating and sex differences, parent-offspring conflict, sibling rivalry, self-deception, taboo, coalitional psychology, and what he calls, following Robert Trivers, the "moral emotions," including sympathy, anger, gratitude, and guilt, which are supposed to modulate reciprocity in a manner compatible with reproductive success.[10] As Brian Boyd, Joseph Carroll, and Jonathan Gottschall insist as they introduce their reader *Evolution, Literature, and Film* (2010), which includes a chapter by Pinker, evolutionary theory of the arts should prevail in the competition amongst theories. Nodding to circularity, a Darwin machine picks the winner in a future where only the fittest theory survives: "Evolutionary theorists of the arts aim not just to offer one more 'school' or 'approach' to fit within the grab bag of current theories, but to alter the paradigm within which studies in art and culture are conducted. We have rallied to Edward O. Wilson's cry for 'consilience' among all the branches of learning. We envision an integrated body of knowledge extending from theories of subatomic particles to theories of the arts. Within this consilient worldview, evolutionary biology is the pivotal discipline uniting the hard sciences with the human sciences and the arts. . . . A true understanding of culture must be rooted in the biological characteristics from which all human cultures grow" (Boyd, Carroll, and Gottschall, 3). Again problematic is the vast distance between the evolutionary-psychological model and the work of literature itself. Pinker, for one, explicitly distances analysis from the local culture of its object: "one has to show—*independently of anything we know about the human behavior in question*—that X, by its intrinsic design, is capable of causing a reproduction-enhancing outcome in an environment like the one in which humans evolved. This analysis can't be a kind of psychology; it must be a kind of engineering—an attempt to lay down the design specs of a system that can accomplish a goal (specifically, a subgoal of reproduction) in a particular world (specifically, the ancestral environment)" (Pinker, "Towards a Consilient Study of Literature," italics in the original, 170).

For their part Boyd, Carroll, and Gottschall emphasize how reduction to more fundamental principles can at the same time reveal new diversity and intricacy, just like reducing heredity to genes and DNA. Granted. But when heredity is reduced to genes and DNA the effect of ecological factors is deemphasized, for instance, and therefore the phenomenon itself—in this case heredity—is oriented toward particular understanding that will better explain something like hemophilia, which is a recessive sex-linked, X chromosome disorder and therefore more common in males, but will not be able to predict how plant clones grown at different elevations will respond (the classic study of *Achillea* discussed by Richard Lewontin in *The Triple Helix: Gene, Organ-*

ism, and Environment). Reductions are required in science whether oriented toward the gene or toward the environment, and for that matter reduction is built into many if not all human projects. But it is disingenuous to imply that reduction may not leave anything out; "new diversity and intricacy" will run one way under one sort of reduction and another on the reduction that runs elsewhere. Decidedly, consilience on the Wilson model runs away from basic paradigms and procedures in the humanities.

So am I accusing Boyd, Carroll, and Gottschall et al. of a category error as they explain it in a section on Frequently Asked Questions? Do they mistakenly want to study art in terms of the sciences? Yes and no. Yes, I agree that "we can reasonably study both human beings and their cultural products using concepts from the human sciences" without making a category error, and in fact the arts including literature have been studied with varying degrees of success according to all sorts of scientific paradigms (for one recent example, see Franco Moretti's Lit Lab and his book *Distant Reading*). At the same time, no: studying art in terms of the sciences cannot proceed according to the reduction noted above without acknowledging the reduction and at what cost; hierarchies of interpretation make a difference as these authors readily acknowledge when they respond to a question about "emergent concepts." Their skeptic asks next: "Doesn't analyzing literature and film require us to invoke concepts and methods different from those in biology and the human sciences?" Like any other subject area, the authors respond, literature has emergent concepts that are "peculiar and appropriate to its own subject," like narrative structure, genre, and point of view. But here's the twist: "All such categories engage attention, imagination, cognition, and emotion in ways that can be most deeply explored in terms of an evolved and culturally developed human nature" (8). Continuing my formal critique of the argument launched above, this fallacy is called the "undistributed middle." Even if we grant (which we shouldn't) that attention, imagination, cognition, and emotion can be *most deeply* explored in terms of an evolved and culturally developed human nature (I think it depends on which aspect of these capacities we study), that does not mean all categories that *engage* attention etc. can be most deeply explored this way. I would suggest, moreover, that these two argumentative fallacies—undistributed middle, and affirming the consequent, which I discussed earlier—are not just nitpicking at the margins of a larger project that may be sound at its core. And this is important: These two instances illustrate deep-seated fallacious reasoning because the basic project cannot otherwise bridge the yawning gap between evolution theory and the literary artifact; fallacies of scale and distance abound precisely where careful step-by-step reasoning is most needed. Regularly we see arguments that launch with a categorical premise that cannot be denied: we should all (even humanists) respect the scientific method and the most current research data; Darwin's theory of evolution by natural selection along with the discovery of particulate gene inheritance (the "Modern Synthesis") is basically right; humans have evolved like all other forms of life; the mind and other basic features of humanity have evolved likewise. Premises granted. But then Literary Darwinists need to plausibly link these granted premises to interesting (i.e., non-vacuous) conclusions, and here's where things get dicey. I may in fact grant all of the premises listed above, I might in fact be "pro-science," without having to conclude that topic X in the arts/literature is most deeply explored in terms of an evolved and culturally developed human nature. Far from it. And that is why fallacies of scale and distance abound, because these critics must regularly move from undeniable science-oriented general premises to close reading of a text while appearing to uphold reasonable standards of argumentation—hence the form of a sound argument without the soundness. The broad outline of this (misleading) arc is typically repeated at the level of book organization. Prototypical is the *Evolution, Literature & Film* reader that starts with Darwin, Dawkins, and E. O. Wilson et al. on the theory of evolution, moves to the basic "Riddle of Art" as a human endeavor, and concludes with interpretations of particular works and topics like "Jealousy in *Othello*," "Slash Fiction and Human Mating Psychology," or "Paleolithic Politics in British Novels of the Longer Nineteenth Century." I should say, by the way, that the movement amongst these first couple of steps is plausible and perfectly interesting; all sorts of worthwhile questions and research projects can and have been built around counterfactual thinking, joint attention, communication, and other basic human capacities shared to a greater and lesser extent with other creatures, which is precisely where scientifically driven research questions appear most appropriate: I think for instance of Michael Tomasello's work across primate and infant cognition. Leaping from this basic order of human capacities to theories of "social change" or to literary criticism, however, fails both on logical grounds and on phenomenological grounds because the analysis must proceed with frameworks and categories inappropriate to the phenomena studied.

In his critical response to Literary Darwinism, Kramnick reminds us a "rapprochement with science" need not be brokered by evolutionary psychology and I would add that any artificial monopoly on the means of interpretation harms (*CI* 433). We need genetics and ecology, and we need literary criticism and scientific psychology, but certainly not on the model of consilience; some alternatives I will explore below and in my epilogue coauthored with ecological neuroscientist Stephanie Preston.

Ostensibly both the new ep and CAL projects would be additive and not subtractive. In the piece, "On the Origin of Comics" repurposed for the *Critical Inquiry* polemic, Boyd explains how "evolution offers a unified and naturalistic causal system from the general to the very particular. Far from reducing all to biology and then to chemistry and physics, it easily and eagerly plugs in more local factors—in a case like comics, historical, technological, social, artistic and individual factors, for instance—the closer we get to particulars. Evolution accepts multilevel explanations, from cells to societies, and the last full room for nature *and* culture, society *and* individuals" (*Critical Inquiry* 395). Likewise Zunshine, again quoting Spolsky, suggests "the addition of biological, evolutionary, and cognitive hypotheses to the discussion of [cultural] change . . . offers literary historical and cultural studies a way to consider the universals of human cognitive processing as they function in their several contexts."[11] Departing from Boyd, however, Zunshine would strike an important note of caution when it comes to how these several contexts add up: "In counting human cognitive architecture as a crucial factor contributing to cultural change, cognitive literary critics subscribe cautiously to a view of this architecture as flexible—cautiously, because although the architecture itself does not change (having remained constant across the species for at least the last 10,000 years), the ways in which it expresses itself in specific environments certainly do" (62). And on this note we are directed to the controversial work of evolutionary psychologists John Tooby and Leda Cosmides (Center for Evolutionary Psychology), who were teachers of Lisa Zunshine when she was an English graduate student at the University of California, Santa Barbara, and who are fellow travelers with Pinker, Trivers, and Buss in the "narrow" or "orthodox" school of EP. In the end, a consilient project of this sort will not be additive in some neutral sense where each of the disciplinary variables carries equal weight, or where critical theory, for instance, might sometimes prevail. "Universals of human cognitive processing" remain constant and will therefore range over the entire series of variables, whereas particular "expressions" will not.

The result of this initial weighting is that the interpretations of any given expression gravitate back toward universals which are, themselves, hardly the object of consensus beyond the highest level of generality like the capacity for language, and especially for Zunshine, Theory of Mind, hereafter ToM (16). From this CAL perspective, no longer is it acceptable to claim some interpretive domain safe from science, and more importantly, no linkage is considered too distant for analytic purposes. The latest sciences of brain physiology, molecular biology, cognitive psychology, and evolutionary psychology should all have a place because their objects of study (e.g., neurons, hormones, language capacity, principles of sexual selection) are each necessary conditions for what appears on the page. No doubt this is true in some trivial sense. But begged are questions of phenomenology: What level of explanation is most appropriate to the phenomenon at issue? J. J. Gibson, another kind of scientific psychologist, puts it this way as he anticipates what is for Jerry Fodor a cynical rhetoric: "The unit you choose for describing the environment depends on the level of the environment you choose to describe."[12] The next argument of this chapter is that Gibson helps us find descriptive units appropriate to human environments not just as they appear in our prehistory, but also as they appear in our literature.

Looking ahead toward my Gibsonian reading of Radcliffe, we can sharpen the point by witnessing how Zunshine treats descriptions of nature that might seem far removed from her cognitive universals, and especially ToM. In *Why We Read Fiction: Theory of Mind and the Novel* (2006), Zunshine recounts a talk on ToM and literature when she was asked about those parts of fictional narratives, like descriptions of nature, that ostensibly have nothing to do with "reporting or guessing character's minds" (25). Zunshine tells us that descriptions of nature are quite scarce, so she looked "in vain" as she returned to Turgenev where, she admits confusingly, nature passages bored her so desperately in her adolescence (she "finally learned to skip all of them").[13]

How does a grown-up Lisa Zunshine train herself not to skip nature passages as she develops into a professor of English literature? Tacking toward humor she "vitiates" her taste with Wordsworth's *Prelude* at the same time she learns to read nature descriptions as pathetic fallacy and personification. Witness how she summarizes this process into One Big Fiction for the sake of her prehistoric cognitive architecture: "If we conceive of the fictional narrative as a cognitive artifact in progress—an ongoing thousand-year-long experimentation with our cognitive adaptations—we can say that this narrative constantly diversifies the ways in which it engages our Theory of Mind" (26–27). Descriptions of nature turn out not to be, in fact, descriptions of nature at all. Or at least they cannot be readily enjoyed as such because irrelevant to ToM. Instead they are mental prompts: "Imagined landscapes, with their pathetic fallacies, personifications, and anthropomorphizing, and with their tacit illuminations of human minds perceiving those landscapes, prompt us to exercise our ToM in a way very different from the stories that contain no such landscapes" (27). While trying to avoid the pathetic fallacy, Zunshine's approach commits a mentalistic fallacy shared by some others in the critical movement CAL. This fallacy reduces anything cognizable—which includes everything written down in a novel—to "mental architecture as defined in cognitive science" where we find four levels according to Patrick

Colm Hogan: intention, representation, connectionism, and neurobiology (*Introduction to Cognitive Cultural Studies*, 237–239).

Yes, nature description in a novel may say something interesting about mental architecture in this technical sense, and the early CAL author Elaine Scarry demonstrates how so.[14] First Scarry makes good use of J. J. Gibson on kinetic occlusion and perceptual acuity that gives us a world—whether found or built—reliable in its vivacity, persistence, and continuity. In the first chapter she shows how, for instance, an author can coax a surface into solidity by passing something transparent over it, using Proust's description of a magic-lantern image playing across the walls at Combray (11). Because the verbal depictions of narrative (as opposed to daydreams or paintings) are almost totally devoid of actual sensory content beyond crosshatches on the page (5), Scarry argues, they imitate perceptual cognition instead of activating perceptual cognition like music, or even a poem where the visual disposition is a prelude to sound (7).[15] "Literature consists of a steady stream of erased imperatives" (35) that dispose us toward a world. By way of demonstration, Scarry reads into the opening of *Tess of the d'Urbervilles* an intricate array of small instructions that include: "[*Look closely at the walker's legs.*] The pair of legs [*now picture their work of weight-bearing*] that carried him [*assess how well they hold that weight*] were rickety, [*and how that affects his motion*] and there was a bias in his gait [*watch which way the load leans*] which inclined him . . ." (37, emphasis original). Referencing Gibson, Scarry says what we notice in narrative, like what we notice otherwise, is in some ways more about the reliable givenness of what we don't notice. The ambulatory and gravitational givenness of this walker in a world of walkers differs from the givenness of creatures who share this world in some ways but not others. Gibson might point out that a young walker walks differently, a tiny creature is subject to gravity differently, and so on. Just as narrative gives us an angle (ambling wherefore) on a shared world that is largely familiar (a world where legs walk), perceptual cognition gives us an angle on a shared world that is largely familiar. Hence for Scarry, narrative imitates perceptual cognition. For Scarry in the spirit of Gibson, narrative is thus a distinctly useful experiment in cognitive psychology as a human endeavor, and at this point I will return to environmental psychology in its most obvious example. For next, Scarry focuses on the flower including Ashbury's delphinium and Rousseau's umbellifer, imaginable for its size that lets it sit in front of our faces and migrate comfortably into our heads, its cup-like shape that breaks over the curve of our eyes, its localization confirmed in the experimental literature of cognitive psychology that explains how smaller images are more "filled in" because they demand less mental energy (53). *Pace* Zunshine, there is no literary problem with nature-out-there boring us until we figure out how it is relevant to Theory of Mind. For Scarry, "nature" like a flower is already part of our mental architecture, and then via criticism, narrative can help us understand how so.

Cognitive Approaches to Literature have been brewing for a few decades, including this rich but relatively unassimilated work by Scarry, along with "cognitive rhetorics" emerging from the 1980 George Lakoff and Mark Johnson publication, *Metaphors We Live By*. One quality that unifies these earlier works is embodied cognition à la Gibson, where embodiment is not just a matter of the living organism defined by its outer membrane, but is rather a matter of environment. For Scarry, we have seen how perceptual cognition involves ambling in a certain way from Shaston to the village of Marlott. For Lakoff, who agrees with Gibson on embodiment but not environmental realism, we are told how categories like anger are inherently "cognitive" insofar as a category is a mental entity, at the same time that it is a culturally defined experience. Hence for Lakoff, affordances of a human environment cannot be invariant as Gibson describes them, as he focuses on certain kinds of physical activity: invariably nature affords locomotion according to Gibson.[16] Instead Lakoff might point out how a remark is inflammatory at a level of detail that varies according to local laws and injurable people; emotional affordances of the human environment are not invariant. So in contrast to Zunshine and recent critics following brain-focused science—Hogan's "mental architecture"—we find a few important distinctions. First, in Scarry and Lakoff one wouldn't even encounter "nature descriptions" unless carefully defined by way of genre (for instance the "nature guide"), because the very topic relies upon misguided commonsense about brain-world division that produces Zunshine's anthropomorphisms and projections at the same time that it obfuscates basic cognitive phenomena like flowers and anger. Second, following Gibson, "environments" for Scarry and Lakoff appear on a spectrum ranging from more or less built to more or less found, and they are shared variously with other conspecifics and potentially, I would add in the spirit of Gibson, with other creatures. Third, in Scarry and Lakoff one sees the investment in scientific inquiry without the red flags of evolutionary psychology. Evolutionary psychology would have nothing interesting to say about Scarry's ambling or Lakoff's inflammatory remark, although both appear under the sign of cognitive science not antiscience.

It should come as no surprise that Zunshine's approach is more controversial than her ecumenism suggests; the interpretation of literature (or anything else for that matter) starts somewhere for some reason and proceeds across fields already populated with people who care one way or another, with invitations and recalcitrance, with paths of greater and lesser resistance up to a vanishing point where opportunities eventually run out. But this is not to say biological science is in principle and completely incommensurable with literary criticism. From the perspective of literary criticism it matters which biological science we are talking about, and in J. J. Gibson's affordance theory we find a point of collaboration where human environments, including nature descriptions, are at issue. The next section explains what is suggestive and limiting in Gibson's affordance and related theory, focusing on the example of a "hostile environment," which we'll encounter in the Gothic novel.

Gibson's Affordances and the Humanities

Roughly where does affordance theory appear at the intersection of science and the humanities? When the philosopher of science Andrea Scarantino gives us a definition wherein "the affordances of the environment . . . are the offerings of the environment, roughly the sets of threats (negative affordances) and promises (positive affordances) that characterize items in the environment relative to organisms,"[17] we in the humanities tend to think first about how threats and promises appear in certain kinds of rhetorical situations, and then, following Scarantino, we might turn toward organisms that include but are not limited to people. Below I will analyze some significant relationships amongst three types of environment where people appear, including 1) an environment that is dangerous for the human-as-organism; 2) an environment that is hostile for people defined according to legally protected classes; and 3) an environment that is hostile in the sense that includes a literary character like Antonia, or La Motte from *The Romance of the Forest*. But first let us recall what threats and promises have meant in the humanities, starting with Aristotle's *Rhetoric*, which analyzes political speech along just these lines. A methodological claim of this section is that the ancient discipline rhetoric is built into affordance theory from the start because that is how threats and promises appear for people.

Here is Aristotle's example. After capturing the island of Samos, Athenians debated sending settlers there in violation of the intent of the Second Athenian League of 377/376. Cydias asks those engaged in a debate about allotment of land to "imagine [all] the Greeks standing around them in a circle, actually seeing and not only later hearing about what they might vote" (1384b32–35). In this case, the rhetorician invokes the figure of a tribunal of conscience in order to evoke shame among those whose reputation is potentially threatened—and this to help the assembly reach a decision compatible with the League's agreement. For Aristotle's rhetorician, threats and promises are speech acts that take narrative form. Moreover in telling this story about a tribunal of conscience, Cydias virtually threatens the debaters: it is a speech act that does something as it is said. Or we can put this in terms of affordance theory: the rhetorician "characterizes" as threat—a negative affordance—items in the environment relative to Greek citizens, where the "relevant items" are community beliefs appearing virtually ("imagine the Greeks seeing . . ."), and the "threat" is a story about damaged reputation. Following rhetoric, then, literary studies appear on the horizon where stories find their home.

But before we turn to literary studies per se let us finish with the pioneer field in the humanities: rhetoricians have long been experts in affordance theory as it pertains to politics and other classical speech situations. Following a famous 1968 article by Lloyd Bitzer, the "rhetorical situation" includes an *exigence* (in terms of my project a basic discomfort[18]) like Congress of Racial Equality advocate Jack Weinberg thrown into a police car on the UC Berkeley campus October 1, 1964, an *audience* like the agitated students who were inspired to action a couple months later by Mario Savio's "put your bodies upon the gears" speech, and a set of *constraints* that are part of the situation because they have the power to constrain decisions and actions needed to modify the exigence: Bitzer lists beliefs, attitudes, documents, facts, traditions, images, interests, motives, and the like (8).[19] Rhetoric, as a humanistic form of affordance theory, posits a situation that is not neutral but is rather "persuasive," threatening and promising with respect to our human being-in-the-world. More precisely rhetorical theory provides a set of awarenesses specially appropriate to human environments: traditionally next to deliberative rhetoric where the goal is audience activation, we find forensic rhetoric where the goal is ultimately judgment, and epideictic rhetoric (ceremonial praise or blame) where the goal is affecting attitudes. Rhetoric also goes into great detail about how the general vectors of threat and promise can be structured for maximum effect: arrange thus and so, deliver not like this but like that. Let us say rhetorical theory provides an historically rich field guide to the threatening and promising environments that are especially human. In turn affordance theory per se attunes us to environmental factors beyond the control of a "good man speaking well," as Cicero famously put it, *vir bonus dicendi peritus*, and in doing so it provides a bridge from exclusively human situations to environments that include the extrahuman. How should we understand the rhetoric of Gibson's affordance theory, where the human organism shades into human being? Tracking intellectual history and keywords, we find Gibson suggestive on this point before he falters on a positivist understanding of language and residual behaviorism that was prevalent mid-century, as I'll explain.

Providing his own intellectual history, Gibson reminds us how, for Gestalt psychologists, we are hailed by things around us:

> The value is clear *on the face of it*, as we say, and thus it has a *physiognomic* quality in the way that the emotions of a man appear *on his face*. To quote from *Principles of Gestalt Psychology* (Koffka, 1935), "Each thing says what it is . . . A fruit says 'Eat me'; water says 'Drink me'; thunder says 'Fear me'; and woman says 'Love me'" (p. 7). These values are vivid and essential features of the experience itself. Koffka did not believe that a meaning of this sort could be explained as a pale context of memory images or an unconscious set of response tendencies. The postbox "invites" the mailing of a letter, the handle "wants to be grasped," and things "tell us what to do with them" (p. 353). Kurt Lewin coined the term *Aufforderungscharakter*, which has been translated as *invitation character* (by J. F. Brown in 1929) and as *valence* (by D. K. Adams in 1931 . . .). The latter term came into general use. *Valences* for Lewin had corresponding *vectors*, which could be represented as arrows pushing the observer toward or away from the object (138).

Then Gibson explains how the concept of affordance is derived from this background in Gestalt psychology, with a crucial difference: "the affordance of something does *not change* as the need of the observer changes. The observer may or may not perceive or attend to the affordance, according to his needs, but the affordance, being invariant, is always there to be perceived" (138–139).

Very good: Gibson decenters first-person experience as the primary criterion for identifying an environmental affordance. But as we saw in Lakoff's objection, and as we discover with respect to human beings, affordances do vary substantially according to situations defined rhetorically, which is not the same thing as experience- or need-orientation. For example, the cop car roof affords Mario Savio, like most adult humans, a stage for speaking—it is *climb-up-able*—and in this regard it is invariant as Gibson would have it. However, the action-event is not defined by mere climbing or speaking-to. Rather it is defined by speaking consequentially, which does not depend upon Savio's personal experience, nor does it reference centrally his needs. Instead the event is defined rhetorically, referencing in narrative form at least the exigence, audience, and constraints like the relevant legal and administrative apparatus, but then also the stuff including a cop car roof that makes this scene not only possible physically but also consequential and persuasive with respect to future activities. Remember it is precisely this surprising stage and its repurposing that gives us the affordance in the first place; it is variable (i.e., not invariant) according to criteria that appear most vividly in narrative form, and the relevant patterns reside substantially in an environment we call cultural[20] or social,[21] depending upon the analytic frame where the first emphasizes communication structures like myths and rituals, and the second emphasizes organizational structures like social stratification. No, a cop car roof is not "always there to be perceived" as a stage for speaking persuasively, and in this regard it is variant. But to the degree that event is now fixed in the historical record, if you will, the cop car roof *is* always there to be perceived as a stage for speaking persuasively and in this regard it is invariant. Like an open mic at a coffee house, or the podium in the US Senate, cop car roofs in certain circumstances suggest themselves as a speaking opportunity thanks, in part, to this historical event with its afterlives in our political and media culture. In certain circumstances cop car roofs now say *plead on me* or *harangue on me.* And you can multiply examples for all sorts of things in the world hailing loudly from some place in the more recent social or cultural world distinct from the world we share with our earliest ancestors. Hence the central question for the theory of affordances is not "whether information is available in ambient light for perceiving them," as Gibson suggests (140), but something closer to the Gestalt psychologists Gibson would supersede, namely how the human environment is structured so as to reveal any particular affordance.

In fact, Kurt Lewin's *A Dynamic Theory of Personality* (1935) starts with a biological environment before moving to an environment that has some distinctly social characteristics. Lewin references the ethologists Jakob von Uexküll on the biological environment characterized as a complex of foods, enemies, means of protection, etc., that offers an organism some set of functional possibilities (77). But focusing on "Environmental Forces in Child Behavior and Development" (70), Lewin turns to agreeable or disagreeable experiences and the special structure of the child's situation.[22] So valences are defined: "Many things attract the child to eating, others to climbing, to grasping, to manipulation, to sucking, to raging at them, etc. These imperative environmental facts—we shall call them valences [*Aufforderungscharakter*]—determine the direction of the behavior. Particularly from the standpoint of dynamics, the valences, their kind (sign), strength, and distribution, must be regarded as among the most important properties of the environment" (73). Interesting discussions ensue regarding the fields of force, *Umweltkräfte* (83), the transformation of the total field, *Umstrukturierung* (84), and embarrassment followed by withdrawal, *Aus-dem-Felde-Gehen* (90). And it is worth mentioning that Gestalt psychologists like Lewin were not alone in the 1930s analyzing human situations in terms other than positivist or behaviorist: Erik Rietveld has recently written about Wittgenstein, for instance, for whom an expert allows himself to be moved by the situation.[23] According to this Gibsonian reading of Wittgenstein, the process of being responsive to affordances is inseparable from the craftsman's concerns, because he perceives a *relevant* solicitation to act: an affordance that matters to him and is experienced as attractive or repulsive (997). And keeping in mind our first topic, we should note the special affective dimension of this process according to Wittgenstein as everyday aesthetic gestures take shape in the context of discontent, disgust, and "discomfort" (*Lectures on Aesthetics*, 1938).

Most important for our project on well-being and its discontents is Lewin's "hopeless situation" (Lewin, 94). He defines it this way: "If a situation becomes *hopeless*, that is, if it becomes as a whole inescapably disagreeable, the child, despairing, *contracts*, physically and psychically, under the vectors coming from all sides and usually attempts to build a wall between himself and the situation. This is expressed both in the typical bodily gesture of despair (crumpling up, covering the eyes with the arms, etc. . . .) and by a sort of encysting of the self: the child becomes obdurate" (94). No doubt Lewin explains how a "state of tension" arises from the physical facts of the environment and the limitations of the child's own abilities: the pencil does not go over the paper as it should. But "still more important are social factors," Lewin insists, especially the authority or power of adults and of other children (97). For his part, Gibson recognizes that "the richest and most elaborate affordances of the environment are provided by other animals and, for us, other people . . . : When touched they touch back, when struck they strike back; in short, they *interact* with the observer and with one another. Behavior affords behavior, and the whole subject matter of psychology and the social sciences can be thought of as an elaboration of this basic fact." Convenient for our purposes he gives the example of political behavior: "all depend on the perceiving of what another person or other persons afford, or sometimes on the misperceiving of it" (135).

Very well. But quickly Gibson retreats back to his basic question at a distance: whether information is available in ambient light, how these "detached objects with topologically closed surfaces" [i.e., other persons] provide stimulus information including speech, pictures, and writing, which are treated as second-order phenomena as I'll discuss below. How something like politics affords, how exactly invitations and threats are characterized in the social field, remains underdeveloped in Gibson despite his debt to Lewin, who showed great sensitivity to the social field and its capacity to make and unmake people.[24] Ultimately the political situation of Aristotle's Athenians cannot appear as such, and cannot be analyzed adequately, on the level of direct perception as Gibson understands it: the situation is constituted by way of political rhetoric poorly understood as speech-information. Nor can the wholly inescapable and disagreeable—Lewin would say hopeless—situation of rape victim Antonia (*The Monk*) appear as such, nor can it be analyzed adequately, on the level of such direct perception. This situation is constituted by way of literary rhetoric poorly understood as speech information, though at the same time it is real enough.

Gibson's theory was adamantly prelinguistic and pre-cultural; like many (but not all) scientific psychologists, he suffered from paleophilia, which marginalized any human activity that isn't shared with our earliest human ancestors. Gibson would blanch at a theory of affordances that included Aristotle's rhetoric defined as "the art of seeing the available means of persuasion in each case," because this definition is at one remove from direct perception, where the means of persuasion typically cannot be seen. Barbed wire is persuasive directly, but rhetoric takes off at just the point where barbed wire and other immediate threats leave off, giving way to means of persuasion that are readily discernible but not immediate. So we in the humanities must depart from Gibson at the same time that we honor his principle that the unit you choose for describing the environment depends on the level of the environment you choose to describe. If we wish to describe the level of the human environment that can give us "a lawsuit waiting to happen" or where workplace hostility potentially resides, then the descriptive unit we choose will derive from a theory of rhetoric that allows for the force of law. At the same time we honor Gibsonian monism that warns against the mistake where we "separate the natural from the artificial as if there were two environments . . . as if there were a world of mental products distinct from the world of material products" (130). At what point did Gibson violate his own monistic principles setting aside supposedly artificial environments like the wishful musings of a personal injury lawyer, or the hostile environment in a Gothic novel?

The core of my Gibson critique aims at his impoverished theory of language and communication, rooted in the mid-twentieth century positivist focus on language as propositional thought, and despite his protests to the contrary (242), the mathematical theory of communication as it was famously composed in 1949 by Shannon and Weaver (see Fig. 3.1.): Words for Gibson provide the experience of things only at secondhand (63). The *purpose* of words, according to Gibson, is to allow for the accumulation of descriptive information about the environment that can be stored and then passed down to future generations. In other words, "language" for Gibson is not a factor in direct perception; it is, like pictures, a matter of indirect perception and

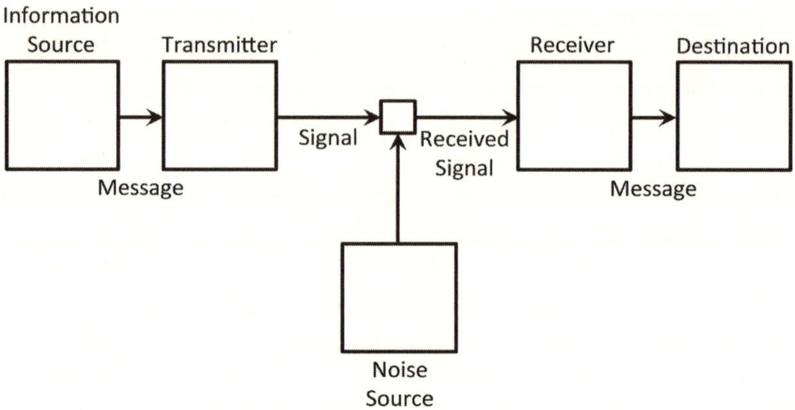

FIGURE 3.1. Shannon Communication System.

hence falls outside his core theory (10, 42) where it forms what he calls "a different kind of knowledge" (253). In this way mind/body, ideal/material dualism returns to Gibson through the back door. However, a practical example helps us rebuild upon Gibson's monistic principles.

Legally considered, a workplace—which is to say a niche defined as a set of affordances (128)—cannot be hostile just because it poses inherent threats that are species-relevant; that would not be a hostile work environment but rather a *dangerous* work environment like a coal mine. Instead a hostile work environment is defined legally with reference to subspecies "characteristics," as they are called in the UK, which are culturally specific: race, age over 40, ability, religion, and so on. Moreover, the "environmental" quality of the hostile work environment is crucial. Actionables are not just hostile behaviors like physical assault or offending words that might be criminal and result in a civil lawsuit directed at the offending person, but also, crucially, nonbehaviors or inactions on the part of an employer. So for example, AT&T had a perfectly good diversity and inclusion policy that might have protected a Kansas City employee and convert to Islam; AT&T did not enforce the policy, and the woman won a hostile work environment harassment case and was awarded $5 million in punitive damages.[25] If AT&T management could have known about the harassment, or failed to take immediate action to stop the harassment, the company could be liable. Unlike nonhuman environments, human environments like the workplace are constituted in part by counterfactuals (e.g., could have known, should have known) that find their grip beyond the handles of immediate perception simply understood.

In a helpful 2003 *Philosophy of Science* article titled "Affordances Explained," Andrea Scarantino indicates differently why Gibson's affordance theory must be revised to account for a range of human environments beyond that which is immediately sensible.

> There is no reason to assume that a theory of perception can explain in the same way how humans perceive, if they do, a number being *divide-by-two-able* (*a mental affordance*), a flying ball being *catch-able* (*a basic physical affordance*), and a flying ball being *score-with-able* (*a nonbasic physical affordance*). Whereas *catch-ability* seems suitable for being grasped in a similar way by language-using and non-language-using creatures, *divide-by-two-ability* and *score-with-ability* demand a story about perception in which the language-mediated ability to master concepts such as *divisible number* or *point in a game* will play a prominent role. This is just to say that affordance properties ought not be treated by default as a homogenous block by theories of perception. They inherit from their constitutive relations with *kinds of doings* and *kinds of happenings* a number of distinguishing properties that are potentially relevant to establish whether they are perceivable, and, if they are, how they are perceived. (960–961, emphasis original)

Scarantino is right that a theory of affordances must go beyond basic physical affordances if it is to include humans. However, his reference to our "language-mediated ability to master concepts" gets us only partway into the hostile environment; it helps us understand how in this case a legal apparatus enables direct perception (of harassing behaviors, for instance). But what should we call the negative affordance we saw in Aristotle's example above, where the threatening environment takes shape precisely as shame posited, the "threat" a story about damaged reputation? A basic physical affordance? A nonbasic physical affordance? A mental affordance? None of these capture the phenomenon adequately. To be precise, this negative affordance does not reference our "language-mediated ability to master concepts" as much as it references the rhetorical situation as defined by Bitzer et al.; and the affordance is not mental or physical, but rather rhetorical as it enacts—or better transacts—a set of relationships that are irreducible to mental entities or physical facts. We might say, referencing Scarantino's terminology, Aristotle's rhetoric provides a "story about perception" without which this situation could not appear as such. Rhetoric, now a subset of language writ large, provides the kinds of doings and the kinds of happenings that make an affordance property like "tribunal of conscience" appear in this situation.

In turn, what could evolutionary psychology say about hostile work environments? Perhaps it could say something general about how people are apparently evolved "to live in groups and to build the cultures they need to survive in a hostile world,"[26] or something about how competition breeds hostility by eroding epistemic conditions like trust and transparency,[27] or how the ingroup-outgroup dynamic shows up all over the social map including in the workplace where the dynamic takes particular historical and cultural form.[28] In other words, evolutionary psychology can say something about hostile environments safety-netted by history and culture, but it cannot say much of interest about the hostile *work* environment per se, let alone a hostile environment à la Radcliffe, and it certainly could not tell us anything about "actionable" events or circumstances that lie at the very heart of the workplace phenomenon after all.

Evocritics have also recently tried to explain what we find inviting and foreboding in representations of the natural world, so we should pay some special attention as we move toward hostile environments in *The Romance of the Forest.* In fact, Stephen Pinker references Gibson in *How the Mind Works* (264) before he moves to aesthetic theory where Gibson wouldn't follow, both because Pinker's analysis depends heavily upon mental representation, and because Pinker slips from the ecology of perception to evolutionary psychology where attractions and repulsion are explained in terms quite alien to Gibson. First Pinker on representation: "People explore a new landscape and draw up a mental resource map, rich in details about water, plants, animals, routes, and shelter. . . . We innately find savannas beautiful, but we also like a landscape that is easy to explore and remember . . ." (376). Then Pinker transitions to evolutionary psychology, exploiting an association fallacy and some nice rhyming at the end of a list. ". . . Geometry of beauty is the visible signal of adaptively valuable objects: safe, food-rich, explorable, learnable habitats, and fertile, healthy dates, mates, and babies" (526). More recently we have Denis Dutton's *The Art Instinct* on "Landscape and Longing" (Oxford University Press, 2009). Like Pinker, Dutton references the famous Orians and Heerwagen study on the savanna as optimal from an early-human survival perspective, and hence beautiful from our aesthetic perspective,[29] also John Tooby and Leda Cosmides on emotion connected to landscape preference (Dutton, 25). "In the Pleistocene, habitat choice was another factor determinative of life and death, and emotional indifference to landscapes is as evolutionarily unlikely as indifference toward snakes, dangerous precipices, and poisonous foods, on the one hand, or sex, babies and sweet and fatty foods, on the other" (26). One example shows how such work is far from Gibson even when Dutton qualifies that the history of landscape painting is about much more than potential Pleistocene habitats (27). When Dutton conjectures that "a climbable tree was a device to escape predators in the Pleistocene, and this life-and-death fact is revealed today in our aesthetic sense for trees (and in children's spontaneous love for climbing them)" (20), he jumps in typical, daredevil fashion from early-human affordance theory to aesthetic theory in our time, and now aesthetic artifacts are appreciated (or not) because they address our species-relevant promises and threats. "Post-Pleistocene signs of settlement or agriculture have become clichés in calendar or greeting-card art perhaps because they seem to humanize a landscape, rendering it less threatening" (20). Or back to Pinker on fiction as one art form that functions just like the others as it exercises our species-defining preferences: "When the illusion works, there is no mystery to the question 'Why do people enjoy fiction?' It is identical to the question 'Why do people enjoy life?' When we are absorbed in a book or a movie, we get to see breathtaking landscapes, hobnob with important people, fall in love with ravishing men and women, protect loved ones, attain impossible goals, and defeat wicked enemies" (539).

Once again my critique of evocriticism as remote suggests itself at the same time a question is begged. I too should say something substantial about hostile environments in the native sense where "hostile" refers to life-threatening enemies and "environment" refers to what we share, roughly, with our prehistoric ancestors. After all, these hostilities are real enough. As we move away from Zunshine's nature description as mere projection, this question becomes important because *The Romance of the Forest* is largely about such primordial threats. I can also frame this issue in terms of a cross-disciplinary imperative. There is no reason we shouldn't be interested in all three of the following: 1) environments dangerous to certain kinds of creatures including us; 2) environments that are hostile to second-class citizens; and 3) environments that are hostile to certain kinds of people defined in rhetorical terms—remember Aristotle's threats and promises that could take shape only with storytelling that references but exceeds the Athenian League's formal agreement. However, these three levels of analysis tend to be distributed to three different disciplinary fields—science, social science, and the humanities respectively—without paying attention to ways in which these three must be interanimate. We do, in fact, have a special discipline beyond rhetoric devoted to storytelling per se, where this third environment gets its due at the same time that it renders social environments and natural environments legible. That discipline is literary studies, where we find David Herman who, invoking Gibson's early contemporary Jakob von Uexküll, cannily calls certain kinds of novelists "*Umwelt* researchers."[30] My argument in the next section is that *Umwelt* researcher Ann Radcliffe composes hostility where it resides at the outskirts of personhood. Or to put this another way, Radcliffe's fiction optimizes the study of hostility because it tells the story with the right social and organic facts.

The Romance of the Forest

It is not "The Romance IN the Forest" but rather "The Romance OF the Forest" that gives us a passionate environment—an environment that will also appear terrible and sublime if genre expectations are realized, even hostile (from the Latin *hostis* or enemy). Next you might think about how, on the back cover of the Oxford edition,[31] "environment" gives way to a passionate "setting": "Set in a Roman Catholic Europe of violent passions and extreme oppression, the novel follows the fate of its heroine Adeline, who is mysteriously placed under the protection of a family fleeing Paris for debt. They take refuge in a ruined abbey in south-eastern France, where sinister relics of the past—a skeleton, a manuscript, and a rusty dagger—are discovered in concealed rooms." (See Fig. 3.2.)

Now for an extended discussion I offer a typical passage, where the debtor and patriarch La Motte scouts a potential refuge. It appears at the beginning of chapter 2 and is preceded by an epigraph from Horace Walpole's *The Mysterious Mother* (1768), which raises in the context of Radcliffe's work the threat of incest, along with some other real threats that appear environmental in Gibson's terms or "circumstantial" in both legal and generic terms; in the last line of the epigraph you'll find my argument in a nutshell.

FIGURE 3.2. *Netley Abbey* (c. 1833), John Constable. (Courtesy of the Herbert F. Johnson Museum, Bequest of David B. Goodstein, Class of 1954)

> How these antique towers and vacant courts
> Chill the suspended soul! Till expectation
> Wears the face of fear: and fear, half ready
> To become devotion, mutters a kind
> Of mental orison, it does not wherefore.
> What a kind of being is circumstance!
>
> HORACE WALPOLE

> He [La Motte] approached, and perceived the Gothic remains of an abbey: it stood on a kind of rude lawn, overshadowed by high and spreading trees, which seemed coeval with the building, and diffused a romantic gloom around. The greater part of the pile appeared to be sinking into ruins, and that, which had withstood the ravages of time, shewed the remaining features of the fabric more awful in decay. The lofty battlements, thickly enwreathed with ivy, were half demolished, and become the residence of birds of prey. (14–15)

As *Umwelt* researchers ourselves, what does this passage teach us about hostile environments, remembering that, for a contemporaneous aesthetic theorist like William Gilpin, "the ruins of abbeys . . . being naturalized to the soil, might indeed, without much impropriety, be classed" as a natural part of the landscape?[32] A few observations in outline before expanding.

1) As opposed to Gibson's enclosure or shelter, La Motte is said to perceive first "the Gothic remains of an Abbey" that was once the "pride of monkish devotion." The abbey is a place built in part for the sake of brutal penitence; at the same time it is half ruined, revealing it "might yet shelter some human being" like La Motte and his crew but also, possibly, some other shady characters like the banditti they had just escaped. In this way counterfactuals are essential to the environment as given. Plausibly the Abbey could be this and it could be that, depending upon the situation that remains carefully undecidable as the narrative unfolds.

2) The high and spreading trees seem gloomily romantic to whom? This is free indirect discourse that intermingles the perspective of author, character, and reader. So yes we might say with a certain Gibson that the high and spreading trees are not real because they are merely invoked by words on a page, which means it is immaterial to ask whose perspective would render them so gloomy. But when it comes to our human environment, can't something be in fact gloomily romantic? And how can we best research an environmental fact such as this? We might set out with *The Sibley Guide to Trees* in hand. No luck.[33] We might track and then interview a few of our goth students on a similar field trip. Better, but not great. Best we carefully read *The Romance of the Forest* and follow up with relevant research in literary and cultural history.

3) Still, aren't these just amusing figments of the author's imagination,

projections onto the natural world, mere anthropomorphism, as Zunshine suggests? To function formally, this complex environment must in fact offer its inhabitants all sorts of trans-species affordances defined "mutually" as Gibson would have it (8). Threatening creatures thereby co-constitute this hostile environment which at the same time affords La Motte and his cohort refuge. Replace "birds of prey" with giant snails on the lofty battlements, and the genre changes completely.

My wager is that a Gothic novel like *The Romance of the Forest* is best equipped to capture this kind of mutuality, which entails precisely the kind of environmental ambivalence where invitation and threat coexist. If these animals and environments are not real at one level of description—La Motte wasn't a living human being and there was no abbey quite like this—the "mutuality of animal and environment," that is, the relationality, *is* real enough, and at the most appropriate level of detail given the rhetorical situation where we choose to describe a hostile environment. An actual La Motte and an actual abbey would by no means afford you better *Umwelt* research if you are interested in hostile and not just dangerous environments. Again recall Gibson: The unit you choose for describing the environment depends on the level of the environment you choose to describe. Now I'll elaborate.

After the first imperfect view of the edifice (Radcliffe, 14), which offers La Motte this initial image not unlike Constable's melancholic scene, the first concrete possibility emerges by way of contrast, as the crew considers how they might survive the night. The carriage had been disabled and could not even "afford a shelter" so after a few moments of silence they return to the most habitable part of the ruin (30). Now settled tolerably—creatures comforted—a new possibility emerges out of the loosely defined structure coming into focus as the days unfold. At that point the narrator tells us for Adeline "the forest, which at first seemed to her a frightful solitude, had lost its terrific aspect; and that edifice, whose half demolished walls and gloomy desolation had struck her mind with the force of melancholy and dismay, was now be held as a domestic asylum, and a safe refuge from the storms of power" (34). Under this description Adeline's domestic wherewithal is exercised for the first time as a quality of her character emerges only in relation: Adeline overlooks and manages "the little affairs of the household" with admirable exactness (34). The very same structure under physical description is some combination of gothic remains, den of thieves, shelter, asylum, and home, each as real as the next, and each revealing to the one who "overlooks" very different exactitudes depending upon the situation. Which subject-object relation prevails in the narrative at any given time says a lot about the character in question and the narrative point.

Asylum—domestic or otherwise—turns out to be a dangerous proposition in itself: because it is granted by the powers that be, and as subject *to*, absolute security is never a given. Security granted through formal asylum only makes explicit a vulnerability we all share as our well-being is dependent upon others. "How had my imagination deceived me!" Adeline says as the La Mottes first appear untrustworthy. "What a picture did it draw of the goodness of the world! And must I then believe that every body is cruel and deceitful? No—let me still be deceived . . ." (150). Self-deception constitutes the basic condition of trust and well-being, or to put this differently we are always vulnerable and achieve some sense of security through our commitments to an environment that includes other people. I try to fit in . . . nothing that bad could happen, could it? This is a special awareness of Gothic fiction: much anxiety and tension rising to the level of terror derives from dependence upon others who may not have one's best interest in mind. Or terror derives from dependence upon institutions like law, convent, and even family that form our initial disadvantages and inevitably betray our trust. Thus also the suspense that characterizes Gothic fiction on the level of epistemology, where certainty is never guaranteed. We rely upon a stranger who may turn out to be a savior or our worst nightmare, upon relations that may turn out to be our closest family and at the same time our undoing, upon stories where credibility seems always poised on the brink of unraveling to the point where what one thinks initially is exactly the opposite of what one discovers later on. The conditions for knowing something are famously unstable in Gothic fiction, "neither here nor there" (145) as the domestic Peter tells stories in a manner that finds some analogy at the level of novelistic structure. Even the surgeon does no better than the domestic. The surgeon is a character at the fringes of science where idealized knowledge becomes in fact life-threatening. "My mind is not of that frivolous kind to be affected by circumstances," says the surgeon just as he botches terribly the treatment of Adeline's injured lover, Theodore (186). What a kind of being is circumstance, indeed.

Nor does a natural-scientific, illusion dispelling, perspective prevail as Radcliffe is read famously by Walter Scott: "all circumstances of her narrative, however mysterious and apparently superhuman, were to be accounted for on natural principles at the winding up of the story."[34] This is the famous "supernatural explained." But is this right? We should pay better attention to the way in which Radcliffe herself frames such explanation. In the midst of reading the horrible manuscript that we later learn is her father's journal-unto-death, for example, Adeline pauses. "A hollow sigh seemed to pass near Adeline and she cried 'Holy Virgin, protect me!', as she threw a fearful glance around the room: 'this is surely something more than fancy'"(134). Murmurings and obscure fleeting figures pursued Emily through the night but when she awoke in the morning, "the cheerful sun-beams played upon the casements, and dispelled illusions of darkness: her mind, soothed and invigorated by sleep, rejected the mystic and turbulent promptings of imagination." Is this supernatural explained simply by fancy exposed to the light of day? Absolutely not, which is what Radcliffe conveys in the very next sentence that reveals a new and even more threatening haunt than the one appearing previously: "this transient gleam of peace fled upon the appearance of the Marquis . . ." (134). Radcliffe's transitions reveal the supernatural not in terms of enlightenment, which has its limits, nor in terms of religion after the fall, but rather in terms of the everyday where real threats take spectral form—slipping through our fingers and evading our glances, just out of earshot—at the same time that they threaten in the most concrete ways. Explaining the supernatural isn't disenchantment as many post-Enlightenment thinkers would have it, but more often in the hands of Radcliffe this is a mere distinguishing gesture that ultimately fails to separate the learned classes from the ignorant, the superstitious from the truly religious, or in the case of Adeline, a character threatened virtually by the stuff of imagination or actually by the Marquis who is also, of course, the stuff of imagination on many levels including on the level of "fiction" that provides the novel its basic conceit. The Marquis is a fiction. But is he? In a formula: real terror, the novel reminds us, is fictional. Terror unlike danger is the stuff of uncertainties as it takes provisional shape and is characterized with respect to our fragile sense of well-being.[35] Hence also the centrality of rhetoric per se, as human being is done up in certain ways.

Later when Adeline falls directly under the power of the Abbey's lawful owner, the Marquis de Montalt, the Marquis is given by Radcliffe a line that turns on the keyword *asylum*, which now appears in all of its ironic force. With designs on Adeline, and La Motte himself against the wall, the Marquis offers Adeline dubious protection: "accept from me the protection he cannot afford" (122). Adeline, at wit's end, throws "a trembling glance upon the prospect around her" and finds that she is at that point practically a subset of La Motte who declares "we are encompassed with dangers" (now remember Lewin's "hopeless situation" where resources are foreclosed). In this case promise = threat: this is precisely the rhetoric of affordances. The Marquis makes a promise to Adeline, an offer "which may afford you at least an honorable asylum" (125). But of course this promise is the worst kind of threat as Radcliffe provides readers an ironic viewpoint they would share with Adeline: abuse of power defines the situation so the affordance is the reverse of what's promised. Instead of honorable, asylum for Adeline would be dishonorable and even deadly as the abbey finally appears as a potential slaughterhouse (338). When early in the habitation La Motte sees pheasants and garden deer—including a friendly fawn—not as neighboring creatures but as "a profusion of game" (23), he also foreshadows how he and his crew will be hunted and ultimately flushed out of their hiding place, poised for slaughter.

Redescription, or the rhetorical figure paradiastole, draws attention to idiomatic expectations negated: on his first encounter with Adeline, La Motte chooses not to abandon Adeline to the "care, or rather, to the neglect" of strangers (12). "Care of strangers" is the idiom that Radcliffe imposes upon her character La Motte by way of free indirect discourse, punctuated in this case by the disjunctive clause set off by commas "or rather," which draws attention to the level at which human environments are basically rhetorical. Moreover Radcliffe is well aware how such alternatives are figured as rhetoric in the technical sense that prevails in that named discipline. Abandoning the sacred rhetoric that should have been her calling, the Lady Abbess is condemned by Adeline: "It was her method . . . to denounce and terrify rather than to persuade and allure" (36). When it comes to human being, threats and promises are fundamentally tied up in language, or recalling Gibson's distinction, they are direct not indirect. More precisely, the threats and promises tied up in language are direct at the level of what J. L. Austin called the perlocutionary (or rhetorical) act, which gets someone to realize or do something like be alarmed or flee. They can be *indirect* at the level of the illocutionary act where, for example, the words mean one thing literally but do in context exactly the opposite. The ruffian says to La Motte "be comforted" and then Radcliffe uses sparingly a graphic device to emphasize irony: "These *comfortable* words renewed the terror of La Motte" (7). So Gibson's direct/indirect distinction can help us understand human environments but not at the threshold of language and other systems that would be supervenient upon our pre-linguistic animality. The distinction can help us distinguish threats and promises at the level of rhetorical analysis where our animality, including our language, sit cunningly askew. "I know not whether I ought to congratulate or console you on this trying occasion," admits M. Verneuil as he twists the plot yet again for Adeline (349), and the reader knows not either because both speech acts—congratulations and consolation—might be appropriate depending upon where one is in the story and which elements merit attention.

At this point, refining Gibson means paying attention to key categorical distinctions already built into our world. Radcliffe shows how affordances are not just itemized threats and promises relative to the organism, nor are they relative to human being, but rather are itemized relative to a particular sort of person or an individual. Most importantly, Gothic horrors impose themselves upon the middling sort including, presumably, the reader along with La Motte and his crew. On one end of the social spectrum we have the domestics like Peter, in whom "curiosity was more prevalent than fear" (19) to the point where he "knew no fear" as his mind becomes wholly occupied with the business at hand (45). Essentially Peter has no emotional intelligence of his own, but instead serves an important function as a vehicle for the "intelligence" he is regularly gathering and transmitting as it is filtered through his own inscrutable designs (49). "Fear" is more for the middling sort like Adeline, who is nonetheless dependent upon someone like Peter if her fear is to be realized. Though Peter seems at times to be lacking his own intelligence, he is a vehicle for the intelligence of certain others: "Adeline waits for Peter, observing his countenance, hoping it might afford some degree of intelligence on the subject of her fears" (138); meanwhile Peter and especially his fellow domestic Annette are constantly retreating from the world that matters, or as Lewin might say, they are characterized precisely in their retreat from the world, *Aus-dem-Felde-Gehen*. At the other end of the social spectrum is the Marquis de Montalt, as removed from domestic service as one can be, but equally fearless. A physician markedly more compassionate than the surgeon worries that Theodore's situation is desperate not only because he is injured but because he appears to be at the mercy of the Marquis, whose emotional life may not be recognizable to someone of the middling sort, at the same time that it is deeply consequential for those in his sphere of influence. "The character of the Marquis is too well known to suffer him either to be loved or respected," observes the physician, "from such a man you have nothing to hope, for he has scarcely anything to fear" (202). In some basic ways everything that happens around the abbey is stamped by this political economy of emotion, where hope and fear in the sense of obligation have unidirectional vectors, with the Marquis in the middle. As Kurt Lewin was aware, affordances have an emotional valence that can be a function of power. At one point Madame La Motte reflects upon how her husband is obligated to the Marquis "whose territory . . . afforded him a shelter from the world" at the same time that it was in the Marquis' power "to betray him into the hands of his enemies" (210); this is Lewin's *Umweltkräfte*. Up to a point, the entire ecosystem Radcliffe gives us in *The Romance of the Forest*—including landscape and its creatures—is shaped by these secular powers.

But the story doesn't end there. From the very outset such secular powers like the Marquis and his territory give way to a religious sensibility that is equally material. After her first night in the abbey, Adeline looks out the window upon an open part of the forest: "it seemed as if Heaven was opening to the view" (22). Indeed, "The scene before her soothed her mind, and exalted her thoughts to the great Author of Nature; she uttered an involuntary prayer: 'Father of good, who made this glorious scene! I resign myself to thy hands . . .'" (22). As the epigraph to chapter 3 has it, external nature is in some sense scripted, but obscurely, just like the civilization imposed by someone like the Marquis: "are not these woods More free from peril than the envious court? Here feel we but the penalty of Adam . . ." (33, citing Shakespeare, *As You Like It* II.1.3–7). Much later arriving at the foot of Montanvert on the way to the earthly paradise that is Leloncourt, Adeline views the astonishing objects around her and says it is "as if we were walking over the ruins of the world, and were the only persons who had survived the wreck" (265), which, we are told in the notes by Chloe Chard, references Thomas Burnet's theory in *Telluris Theoria Sacra* (Amsterdam, 1694) that mountains were not created directly by God, but were formed as a product of the Flood and constitute "ruins" of previous natural formations (390n265). At Leloncourt, La Luc articulates this theo-geology at the level of concrete being. Though we try to comprehend the sublimity of the Deity who first "called us into being," we fail in part, but can still grasp at that portion of nature communicated. Recalling Lisa Zunshine, wouldn't religious illusions such as this count as mere mental projection, anthropomorphism, and pathetic fallacy? As if to anticipate such skepticism, La Luc responds, "Call them not the illusions of a visionary brain" (274) and we are tempted, in the spirit of Gibson, to apply this injunction across Radcliffe's entire oeuvre. For people of this sort, the threats and promises of a world that includes secular powers may be just as real as the threats and promises of the world that includes nonsecular powers, whether environmental in Gibson's sense, or religious in the Christian sense voiced by La Luc. "Seems" and "appears" are thus crucial devices for the author Radcliffe, who works precisely at the point of indeterminacy where concrete possibilities meet; feeling but the penalty of Adam means that's all we can do. Thank God. At the end of the story, does it turn out that Adeline was an instrument of God all along? That appears to be the case (but who's to say). Why, asks the narrator, did the Marquis dispose of his brother but not his infant niece, Adeline, who stood between the Marquis and his wishes? That oversight "seems somewhat surprising," the narrator concedes on the realism-of-motives principle, "unless we admit that a destiny hung over him on this occasion, and that she was suffered to live as an instrument to punish the murder of her parent" (343). In fact the book ends not with the supernatural explained; it ends more or less with a trial and the king's justice. But that resolution is cut short by the Marquis' suicide and this sense of divine justice that remains a viable alternative, an ever-present but never fully realized opportunity for *Umstrukturierung* as Lewin might say: transformation of the total field.

Now to conclude this chapter. If you want to research the hostile work environment you need to know about the law and some of the relevant cases. But what if you want to know about other sorts of human environments, including hostile, romantic, awkward, thrilling, and so on? Sometimes the ecological concerns of natural and social scientists might in fact find their most inviting home in the wilds of literature.

4

Mixed Feelings in Jane Austen's *Sense and Sensibility*

Mixed feelings are at the heart of Jane Austen's novel *Sense and Sensibility*, and we think we know what that means, like a recipe: combine two parts a feeling like gratitude, one part happiness, a dash of resentment and you get something like Elinor. But mixed feelings in the novel and beyond are poorly served by this disequilibrium model recalling the long tradition back to Galen.[1] In fact, mixed feelings are a matter of negotiated circumstances where feelings may be at odds as they converge on character. Hence the significance of literature and particularly the sentimental novel as a cross-disciplinary research domain, where this kind of rhetorical situation is exquisitely detailed. Referencing the science of situated emotion (Barsalou, Barrett) and behavioral economics (Kahneman on bias and persuasion), this chapter offers a new way to understand mixed feelings as rhetorically situated. At the same time, it explores a methodological opportunity at the interface of science and the humanities, beyond recent work in Cognitive Approaches to Literature, which frequently proceeds unecologically toward theory of mind.

Perhaps it is not surprising that Jane Austen's *Sense and Sensibility* hinges on mixed feelings. After all, its title foregrounds the arc of two characters—a more sensible Elinor and sister Marianne—as they mutually modify toward a happy ending. Paying homage to Austen's utilitarian concern that I'll discuss below, we calculate their feelings mixed at the novel's midpoint between emotional states more stable.[2] In our day such a midpoint calculation might even be realized visually with blending algorithms like this one from the 2005 *Virtual Reality* article, "Mixed Feelings: Expression of Nonbasic Emotions in a Muscle-Based Talking Head" (Fig. 4.1). What is creepy about these images? I think that the blended images in the middle are especially uncanny because

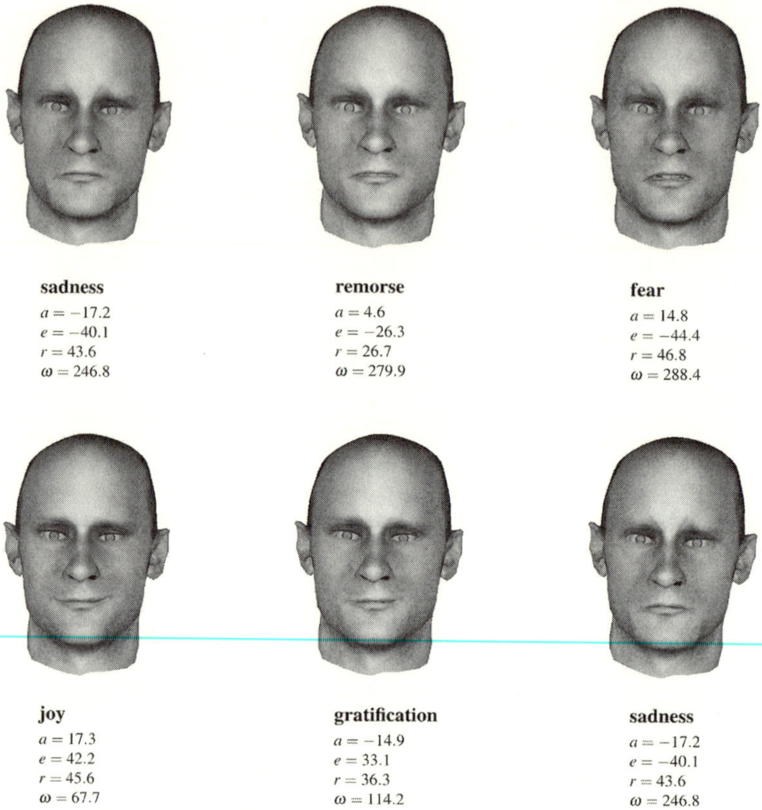

FIGURE 4.1. "The emotional expression in the middle has been obtained from those at the left and right using the blending algorithm. The radius *r* and the angle on the emotion disc *x* determine the influence of each generating expression and hence the degree of similarity to the new one. The first example is a not too active, but rather negative emotion, while the second one could be pleasant surprise, and the last one unpleasant surprise." (Reproduced with permission from Springer Nature)

they depict emotions mathematically unhinged even from implied humanity.[3] But more on the creepy side of math later.

Returning to Austen, we might alternatively figure emotional stability in terms of a trial-filled journey as the novel opens with the Dashwood women happily "settled" in Sussex (3) and ends with them happily *re*settled in Barton and Delaford where "strong family affection" is easily communicated (289).[4] Generally speaking, the homeostatic formula EMOTIONAL STABILITY→ MIXED FEELINGS→ EMOTIONAL STABILITY might characterize the novel, its central characters, and even a version of our own folk psychology that would locate basic emotions in the person and then treat their mixture as some kind of measurable midpoint[5]: "I FEEL NOT HALF BAD." "I HAVE HALF A MIND TO . . ." where the dot dot dot of aposiopesis leads to nothing good, while the equivalent German idiom, "Ich hätte fast Lust zu . . ." confirms that the half mind we are talking about in this case is indeed the emotional half. "I FEEL TORN," in half we can presume, if Gilles Fauconnier and Mark Turner were right about conceptual blending.[6] To summarize, we still live in the wake of a general utilitarianism famously articulated by Jeremy Bentham just before Austen drafted in the 1790s a precursor epistolary novel "Elinor and Marianne." For the crass utilitarian, emotion, and most importantly happiness, can be quantified and therefore halved.[7]

But who wants to be a crass utilitarian? Certainly not Jeremy Bentham, as I will note below. Then which alternatives do we have, beyond Coleridge?[8] Fewer than one might think, because the fog of pre-calculation hangs heavily, as we will see. However, with the help of Jane Austen and some unlikely allies in our contemporary psychological and cognitive sciences, I will pose in this chapter an alternative where the rhetorical situation is central—more precisely, I will explore how mixed feelings materialize in reference values that are negotiated only by way of rhetoric. Now back to our muscle-based talking heads.

Blending algorithms no doubt have their application. Art is transformed by the aesthetics of a mathematical procedure that gives us Pixar, and feelings by a newly reliable form of expression that would allow us to track their mixed physiognomy back to basics. "Basic emotions" are again a technical designation defended by evolutionary psychologist Paul Ekman. These are emotions evolved for their adaptive value in dealing with fundamental life tasks such as achievements, losses, frustrations, and threats, where the paradigm is ingesting something repulsive as a cause for disgust, or ingesting something attractive as a cause for enjoyment (see chapter 1 for images and discussion). Take "surprise," which we will revisit in Austen by way of astonishment. For Ekman, the term "basic" postulates among other things "that other non-basic emotions are combinations of the basic emotions, which may be called blends or mixed emotional states," for example, smug = enjoyment + contempt, though he offers the qualification that what we perceive as mixed emotions on the face are likely basic emotions passing in microseconds.[9] In comparison to a basic emotions approach, Austen's emotional rhetoric will seem quaint, wordy, and relatively obscure. Does that mean that the rhetoric of emotion is dead, at least from a scientific perspective? Perhaps not, and in fact the rhetoric of emotion may be newly evident if we return to Austen from the perspective of our post-utilitarian moment, which I will explain below.

The key *Sense and Sensibility* scene comes in the final Volume III when Elinor, in reciprocal love with unavailable Edward, is obliged to inform her beloved that his languishing engagement to rival Lucy might now be realized since Colonel Brandon has promised him a small rectory with income enough to support a modest marriage. As Elinor prepares to write Edward, he appears suddenly on the scene. For the second time in so many chapters Elinor is astonished[10] and feels "particularly uncomfortable for some minutes" (217) before she delivers the news with congratulations, reporting with some irony the £200 per annum might just establish all his "views of happiness" (218), including, it is presumed, his marriage to Lucy. The narrator tells us, "What Edward felt, as he could not say it himself, it cannot be expected that anyone else should say for him. He *looked* all the astonishment which such unexpected, such unthought-of information could not fail of exciting; but he said only these two words: 'Colonel Brandon!'"

So in this most emotional scene, our crass utilitarian efforts are frustrated on at least three levels. First, contra Lisa Zunshine, Blakey Vermeule, and like-minded literary critics (see previous chapter), mind reading per se is frustrated insofar as Edward's legibility is denied not only in terms of external expression, but also in terms of feeling that the narrator tells us Edward could not say to himself. This would make any pre-calculation of happiness or of any other emotion difficult, to say the least. Second, contra calculating externalists, including the character Marianne, who would look to the face for physiognomic truths, Edward's emotional expression refers only to absence at the heart, but not apathy. A physiognomist sees on Edward's face only "astonishment" (Fig. 4.2) that registers, in its unadulterated form, nothing but change per se, or what the narrator calls "excitement." Like our muscle-based talking head, canonic representations of astonishment[11] incorporate the scalar inflection admiration and fear, suggesting that it is difficult to picture astonishment per se. In fact the 1833 *Pride and Prejudice* "astonishment" illustration relies not primarily upon physiognomy, despite Mr. Bennet's bugging eyes, but fundamentally upon context, including narrative reference points that situate Mr. Bennet's astonishment at the awkward moment when trepidation about Darcy is no longer viable because he has saved Elizabeth's sister Lydia, though a sentimental resolution is still nowhere in sight. Indeed this moment is awkward in the most profound sense because it destabilizes the emotional arc in both temporal directions as the focal characters, and we as readers, must revise feelings already felt—since prejudice against Darcy's "pride" turns out to be unjustified. Another way to put this is that the caption referencing in turn the narrative is essential. Astonishment—or what we now call surprise[12]—renders any calculation of happiness or another basic emotion difficult if not impossible since it is only a common denominator that registers: fear over surprise, admiration over surprise, anger over surprise, and so on through the six "brief and episodic" emotions that Ekman distinguishes from enduring emotions like love or boredom that he prefers to call "affective commitments."[13] (See Fig. 4.3.)

In fact, scientists of emotion have always confronted this problem in the laboratory, since fMRI or other measures of emotional response must account for the response phenomenon itself; that response phenomenon historically has registered as attention or "excitement" or indeed "surprise." For instance, the eminent psychologist Jerome Kagan cites one gender study

FIGURE 4.2. "She then told him what Mr. Darcy had voluntarily done for Lydia. He heard her with astonishment." Illustration from *Pride and Prejudice* by Jane Austen. Engraved by William Greatbatch 1833. Mr. Bennet is on the left, Elizabeth on the right.

FIGURE 4.3. [Left] Charles Le Brun, Expression Studies: Astonishment with Fear. Black chalk, pen and black ink on paper. (©Musée du Louvre, Dist. RMN-Grand Palais / Martine Beck-Coppola / Art Resource, reproduced with permission) [Right] Admiration with astonishment. Red/brown ink. Stipple Engraving 1785 after Charles Le Brun. From Bowles's *Passions of the Soul.* (Wellcome Library, London, CC BY 4.0)

where subjects judging facial expressions in a laboratory took longer to decide whether a man's face was fearful because they did not expect to see photos of fearful men (63). Generally Kagan urges us to consider relations between an emotional incentive and a brain profile—for example, a fearful face photograph and the viewing subject's amygdala activity—in terms of response uncertainty or surprise, not just in terms of incentive-response matching where a fearful face maps directly onto a fearful subject response.[14] Likewise in our Austen passage, astonishment obscures precisely because it is reflexive and all too predictable; it is a reaction "such unthought-of information *could not fail* of exciting."

So here is the question. Would a God's eye perspective, or an advanced measurement system, or airtight evolutionary psychology allow us to resolve or read Edward's mixed feelings as an implied human being,[15] if not a real one? No, according to Austen, who finally frustrates this fantasy by withholding omniscient narration at the crucial juncture. But then aren't we left only with skepticism and nothing at all to say about mixed feelings, undermining both our analysis and perhaps more importantly the story itself? The first anti-skeptical clue should come from the simple fact that Austen's story does proceed, and quite successfully if our continued interest and reception history is to be any indication; even philosophers like Gilbert Ryle and Alister

McIntyre have read Jane Austen as a narrator of emotional equilibrium.[16] So how does the story proceed? By way of mixed feelings understood in contrast to the Ekman program of basic emotions. Now I will briefly turn to this contrastive understanding as it has emerged recently in cognitive science research.

In their important critique of the basic emotion program culminating in the 2011 *Neuropsychologia* article "Grounding Emotion in Situated Conceptualization," Lawrence Barsalou, Lisa Feldman Barrett, and their colleagues render Kagan's uncertainty problem productive by first separating with catch trials the response to an emotional incentive and what they call its *conceptualization* (1112).[17] Take a look (Fig. 4.4) at Sen. Jim Webb's face in (a) and name what emotion you see. Most say anger, or something similar. Now take a look at the larger context (b) and name what emotion you see. Most say "elation" or something similar. Instrument- and perception-based findings frame what these researchers call the "emotion paradox" where people can automatically see Jim Webb as either angry or elated, depending upon the image, even though sufficient information for this judgment is not unambiguously displayed on his face or body.[18] How can we do this? Because, Barrett and Barsalou argue, emotions are not brain mechanisms triggered in a stereotyped and obligatory way as if a coiled snake triggers the fear circuit instinctually ("Grounding Emotion," 1105). Instead emotions are fundamentally situated, and we have a sense for how these situations work even

FIGURE 4.4. "Look at United States Senator Jim Webb in (a). Taken out of context, he looks agitated and aggressive. Yet look at him again in (b). When situated, he appears happy and excited. Without context, and with only the structural information from the face as a guide, it is easy to mistake the emotion that you see in another person. A similar error in perception was said to have cost Howard Dean the opportunity to run for President of the United States in 2004." (Original photograph by Doug Mills/The New York Times/Redux. Reprinted as formatted in *Trends in Cognitive Sciences*, Vol. 11, No. 8, edited by Lisa Feldman Barrett, Kristen A. Lindquist, and Maria Gendron, "Language as Context for the Perception of Emotion," p. 328, 2007, with permission from Elsevier. Original photo reproduced with permission from Redux Pictures.)

when *the situations themselves* are ambiguous. Essentially Barrett and Barsalou want to test in a laboratory this lucid passage from Knight Dunlap's 1932 article on situated emotion: "I may apprehend my situation primarily as one in which my welfare is threatened. I call my inner responses *fear. You* may apprehend my situation as one in which I fight against the threat; you say I am *angry.* Which is correct? The answer may be: Both!"[19] Now you may wonder about the brain physiology in this case. Would a brain scan of Dunlap's subject reveal significant differences if the situation was associated with the emotional concept *fear* as opposed to *anger*? Barrett and Barsalou say yes, and here is their account.

Fear of a runner who becomes lost on a wooded trail at dusk differs from the fear of someone unprepared to give an important presentation at work because the agent conceptualizes the situations differently ("Grounding Emotion," 1108) and therefore mobilizes different aspects of the emotional brain—in the case of social fear, clusters are mobilized in the left dorsolateral prefrontal cortex and inferior frontal gyrus, which are together associated with executive control, not motor response. In other words, social fear gets our wheels spinning, physical fear gets our legs moving (1124). Moreover two *different* emotions—in this case fear and anger—produce significantly different brain images not because these emotions evolved for their adaptive value in dealing with different life tasks à la Ekman, but because their situated conceptualization differs. So how does the experiment proceed? In a manner that speaks to the literary humanities and especially the art of rhetoric.

> In an initial training phase, participants became familiar with two situation types. Importantly, these situations were constructed so that a participant could experience either anger or fear within the context created. . . . On two separate days before the critical scans, participants listened to situations of each type and rated each situation for familiarity, imagery, and their ability to "be there" (i.e., immerse oneself in the situation). As participants listened to a situation, they were instructed to immerse themselves in it as deeply as possible. Descriptions of the situations were written from the first person perspective and contained various details designed to induce immersion. ("Grounding Emotion," 1110)

In fact when we look at an example below, we will see that descriptions of the situations were written from the *second*-person perspective, which begs literary and rhetorical questions about identification and the ability to "be there," or immerse oneself. And although they are careful to explain how "concepts" are Wittgensteinian without a core (e.g., a "game" has no core though we can point to prototypes, 1106), Barrett and Barsalou could learn more from humanistic philosophers of situated cognition who explore how emotions function not conceptually by way of representation, but practically.[20] Then they continue:

> On critical trials during scanning, participants first listened to one of 30 physical danger or to one of 30 social evaluation core situations mixed randomly together. Following the situation, participants heard the word for one of four concepts, again mixed randomly: *anger*, *fear*, *observe*, or *plan*. Participants' task on hearing the concept word was to rate how easily they experience the concept in the given situation. ("Grounding Emotion," 1110) [The two non-emotion abstract concepts "observe" and "plan" were included for comparison purposes.]

Comparing data, the research team concludes that specific instances of emotion are constructed dynamically within neural circuitry distributed across the brain to represent an emotion in a particular situation (1125).

The neurophysiology of mixed feelings is for Barrett and Barsalou situational, and "situation" is defined by familiar narrative and rhetorical qualities. Protocols for writing these mini-narratives are telling, as they may remind you of Cicero's courtroom vivacity[21] or Creative Writing 101, which typically has an assignment that goes something like this: *Describe with sensory detail a setting and activity performed by someone with relevant personal attributes; describe an action, consequence, and resulting somatosensory experience.*[22] Here is an example of a social evaluation situation that practically places you in a scene from *Sense and Sensibility*, where awkward situations abound. Imagine this scenario you may remember from my introduction:

> You're at a dinner party with friends. A debate about a contentious issue arises that gets everyone at the table talking. You alone bravely defend the unpopular view. Your comments are met with sudden uncomfortable silence. Your friends are looking down at their plates, avoiding eye contact with you. You feel your chest tighten.

In this situation, again, would I be angry at myself for going out on a limb, or fearful that my friends might distance themselves further? That would depend upon the situation, which would need more details including my relationship to interested onlookers and the view I defended. Now returning to Austen's critique of utilitarianism, my mixed feelings would be determined not by any predictable calculation but rather in dynamic relationship with onlookers who might in fact be *extrinsic*. Let me explain.

If you recall in Edward's astonishment scene, our utilitarian efforts as *readers* are explicitly frustrated. Although Elinor is implicated by the "look" only she could grant at that moment, "anyone"—including the reader—is denied the insight that would come with an emotional label; Elinor and then we, as happiness calculators, are disabled. Remember the entire scene is framed by a mockery of utilitarian calculation that would have *all* Edward's views of happiness realized by a sum lying somewhere between £200 and $x =$ the availability of Colonel Brandon's servants, carriage, cows, and poultry, as "secretly" calculated by Lucy in the following scene (221). In light of Barrett and Barsalou, we can say that Edward's mixed feelings are defined by interested onlookers who, in this case, include us.[23]

Disingenuously (198), Elinor earlier explains the situation to Marianne in terms of what we now call "expected utility" that rationally calculates absolute value: "'Edward will marry Lucy; he will marry a woman superior in person and understanding to half her sex; and time and habit will teach him to forget that he ever thought another superior to *her*.'" Looking ahead to Kahneman and Tversky, we observe that Marianne offers Elinor a critique of expected utility by addressing Elinor as a non-abstract and unaveraged person who should be interested not in absolute value, but in her own *relative* superiority to Lucy. "'If such is your way of thinking,'" replies Marianne, "'if the loss of what is most valued is so easily to be made up by something else, your resolution, your self-command, are, perhaps, a little less to be wondered at.'"

Not *no* feelings, to use the double negative Austen found so stylistically congenial. Mixed feelings that tempt us to read along with astonishment, excitement, gratitude, and happiness, perhaps also love, regret, shame, unhappiness, disgust, terror, melancholy, and we could go on like Elinor who finally sits down to "reconsider the past, recall the words and endeavour to comprehend all the feelings of Edward; and, of course, to reflect on her own with discontent" (219). Thus our analysis proceeds not despite, but precisely because of the mixed feelings that Austen treats masterfully when considered from this perspective of recent cognitive science and post-utilitarian psychology, including the influential Daniel Kahneman article to which we will now turn: "Reference Points, Anchors, Norms, and Mixed Feelings."[24]

Explicitly for Kahneman, mixed feelings are not calculable intermediates but rather ambivalencies that reference multiple points subject to what he calls, depending upon the example, negotiation or persuasion. Let's see how this works. Kahneman asks us to

> Consider a man who was given good reason to expect a $5000 raise, but eventually received only $3000. What is the psychological value of that event, as the individual experiences it? Obviously, the raise can be coded either as a gain or as a loss, depending on whether it is compared to zero or to $5000. A simple model is that the raise of $3000 is compared to a single reference point, which is a weighted composite of the two salient reference points. This idea is rather

> unappealing, however, because it implies that there exists a truly neutral raise, somewhere between zero and $5000, such that all larger raises are coded consistently as gains and all smaller raises as losses. A formulation that may be more satisfying is that any raise of less than $5000 is experienced *both* as a gain and as a loss, at different times. (306)

Practically, mixed feelings in judgment tasks are irresolvable with reference to single weighted composites or an averaging formula, or a fixed temporality; Kahneman would not be impressed by our blending algorithm, for instance. Nor would be Austen, who might poke fun at a crass utilitarian telling the $3000 raise recipient the following: "Why don't you just add up your gratitude and resentment, elation and disappointment, happiness and unhappiness, then divide by two . . . now feel that way."

So what is Kahneman's post-utilitarian alternative? Mixed feelings with *multiple* reference points. And I note Jeremy Bentham himself qualified any model of calculation that might deliver absolute values for happiness; he was no crass utilitarian of the sort we might imagine along with Austen. Instead, Bentham was deeply interested in what he called "circumstances influencing sensibility," such as gender, education, and rank, which inevitably bias happiness calculation because they determine the scope of one's life. "Cæteris Paribus," speculates Bentham, "the quantum of sensibility appears to be greater in the higher ranks of men than in the lower," greater in women than men, and so on through a list of biases we recognize as historically specific. Bentham's entire project redesigns law and government around biases that would be most advantageous from the perspective of the state.[25]

Meanwhile, here is how biases work for Kahneman. In the example above, the subject references at least the starting salary in comparison to which the $3000 raise makes him happy, and the "expectation" in comparison to which the raise makes him unhappy.[26] It is fully in line with Kahneman's complex approach to also consider the subject's knowledge of comparable situations (i.e., "What's that guy making?"), the firm's balance sheet, and other reference values which in this imagined scenario must have been communicated in such a way that produced unfulfilled expectations. Kahneman makes the point that negotiators, politicians, and advertisers regularly communicate to establish reference points that serve their purposes, where that could mean for instance a labor union negotiator lowballing to reduce expectations, a politician announcing small initiatives with great fanfare, or an advertiser exploiting what Kahneman calls elsewhere the "focusing illusion" that encourages a consumer to believe that a change of weight, scent, or whatever will produce marked improvements in his or her happiness.[27] In a nutshell, Kahneman ties value to a rhetoric of possibilities.

Of course, these are all examples of rhetoric in the bad sense of Plato's sophistry where subjects are intentionally misled to the disadvantage of at least themselves; at points we can see the rhetorical figure paradiastole at work when the truth is opposed by false terms, to invoke Renaissance rhetorician Henry Peacham. But according to Kahneman, persuasion also has its positive ethics; he explains for instance how "a skillful negotiator may be able to smooth the sharp edges of loss aversion and outrage in the views of the other side, by introducing alternative reference points" (307). "I am not losing a vehicle, I am gaining freedom from debt," recommends the Bankruptcy Law Network as they troll for customers. "Remember you are not losing a daughter—you are gaining a son!" To summarize communication, or persuasion, or rhetoric is inillimitable because reference points are not fixed absolutely and they are subject to revision. But against Quentin Skinner's generic account of paradiastole, where "it is always possible to argue in *utramque partem* on either side of the case,"[28] Kahneman's example shows how reference values are lodged in a recalcitrant world subject, however, to negotiation.

In our $3000 raise example, mixed feelings are also inillimitable. But in order to identify, analyze, and explain mixed feelings, we can't just ask this subject how he feels. We need to establish the relevant reference points that produce, according to Kahneman's rather unsatisfying analogy, a duck/rabbit effect for the subject where happiness and unhappiness are experienced intermittently (and thus can be frequency normed): post- not pre-calculated, depending upon which reference point is salient at a given moment.[29] Feelings in this example are mixed, in other words, by way of concrete reference points that can be incorporated into a quantitative model. But the model is also *qualitative* at its core; by way of Kahneman's "expectation," rhetoric constitutes the finite set of relationships that makes feeling quantifiable in the first place (e.g., unhappiness with respect to the $5000 expectation).

Kahneman's point is not that utilitarian models are inherently misleading. Instead his point is that any good utilitarian model must ultimately account for the complex contours of psychology and the communication strategies, or rhetorics, that shape them. And though I'll not make the case here in detail, behavioral economics emerged at least in part by way of rediscovering the rhetoric of economics we find in the likes of Adam Smith, who in the latter part of the eighteenth century integrated his work on rhetoric, moral sentiments, and the wealth of nations.[30] Rhetoric has long been an expert domain when it comes to emotion or so-called irrationalism, and now that irrationalism in the social sciences has once again become interesting, rhetorical studies have a new appeal.

Irrationalism also returns us finally to poor Edward and Elinor, although their situation demonstrates how irrationalism can make all the sense in the world. Last we saw them, Edward was feeling mixed in a manner that was obscured by astonishment. Following Kahneman, let's call him the subject faced with multiple and conflicting reference points, including at least 1) his gratitude toward Colonel Brandon and Elinor, who have cooperatively enabled his minimal happiness; 2) his resentment toward Colonel Brandon and Elinor, who have at the same time *dis*abled his optimal happiness; 3) his sense of minimal obligation toward Lucy, whom he had engaged; 4) his sense of maximal obligation toward Elinor, whom he had more recently wooed; 5) his disgust with whomever, including himself, had put him in this awkward situation; and 6) his specific astonishment that Elinor must also populate this last category because she had thrown their awkward situation directly in his face.

In turn, Elinor might be called by Kahneman an interested negotiator, and we might call her a skilled rhetorician, even a very, very subtle seducer. If you think back over 1 through 6 Elinor has a hand in each: she has extracted Edward's gratitude without allowing that he would owe her anything, she has confronted Edward with an obvious choice between lowered and raised expectations, she has exploited what Kahneman and Tversky famously call *loss aversion* by presenting herself as the more immediate love object at the same time that she articulates the very alienation that would force Edward's hand. No wonder Edward is not just specifically but also generally astonished.

But remember, Kahneman would point out how mixed feelings depend upon reference points outside the scope of what a negotiator controls, indeed outside the scope of any interpersonal situation. The stingy boss may strategically reference a disappointing balance sheet in contract negotiations, but at a certain point the balance sheet is recalcitrant—it is what it is. And of course with Austen we are merely reading a work of literary fiction whereas in Kahneman's example we are . . . well, I guess we are also reading a work of literary fiction, in this case an imaginative mini-narrative populated not with real human beings but implied human beings for whom context never disappears completely into the depths of psyche. There is a way in which scenario building—even for oneself—always relies upon narrative fictionality.

When it comes to the phenomenon of mixed feelings, context matters fundamentally, which makes Kahneman's social science and Austen's fiction different but compatibly insightful. Although the embedding genres are different, our two examples share this question: What is relevant context for the mixed feelings of implied human beings, in this case Elinor and Edward? Paul

Goring's *The Rhetoric of Sensibility in Eighteenth-Century Culture* reminds us how this question goes beyond a particular Austen novel interested in utilitarianism, to at least one of its embedding genres, the sentimental novel. Referencing Samuel Richardson's *Pamela*, Sarah Fielding's *The Adventures of David Simple*, Henry McKenzie's *The Man of Feeling*, and Laurence Sterne's *Sentimental Journey*, Goring demonstrates how in sentimental fiction, the rhetoric of emotion depends fundamentally not just upon the conventions of emotional expression à la Lavater (in our day Ekman) but rather upon scene mechanics, including the staging of witnessed eloquence that implicates a reader. "In relation to both male and female characters, then, sentimental fiction insists upon an association between visible emotionalism and modern virtue, but the actual *eloquence* of bodies represented in fiction—the basis of their emotive hold over readers, and thus the basis of their social potential—hangs not only upon their embodiment of eighteenth-century thinking about the passions' visibility, but also upon the mechanics of the scenes in which they are depicted" (153).[31] With Kahneman we conclude that the emotional "mechanics of the scene" in sentimental fiction materializes in reference values that are negotiated only by way of rhetoric. And this should be no great surprise considering that sentimental fiction, including its *Sense and Sensibility* transformation, was deeply concerned with the interference between emotional economies and economies that would be exclusively financial.[32]

By way of conclusion I will focus on the important role played by gossips John Dashwood and Mrs. Jennings, who allow Austen to establish crucial reference points—indeed we can call them Kahnemanite norms—outside the immediate purview of the central characters. Epigraphically I offer this amusing image from Lisa Feldman Barrett's recent article on the visual impact of gossip where she is interested in how "top-down" affective information acquired through gossip influences vision (Fig. 4.5). Kahneman and Miller's 1986 "Norm Theory: Comparing Reality to Its Alternatives" cites Barsalou on the way to rejecting the treatment of normality based on formal or informal conceptions of probability, offering instead what they call a "phenomenology of surprise" (137). For instance, the PERCEIVED ABNORMALITY of a victim's fate increases sympathy: subjects felt more sympathy for an imaginary victim who was severely injured during a robbery that took place in a store to which the victim had decided to go only after finding his regular store closed for renovations (146). Also the AVAILABILITY OF COUNTERFACTUALS increases sympathy: someone who misses a plane by 5 minutes receives more than someone who misses the plane by 30 minutes because it's easier to imagine the world where 5 minutes not 30 had been shaved off (144).[33] On the other hand, crime victims lose sympathy inadvertently because we tend to consider the ACTIONS OF A FOCAL INDIVIDUAL mutable in comparison to background actors, who would be in this case the perpetrators; hence "blame the victim" (144). Generally, Kahneman and Miller confirm the correlation between the perception of abnormality of an event and "the intensity of the affective reaction to it, whether the affective reaction be one of regret, horror, or outrage" (146). We might say that norms have a rhetoric that cannot be precalculated or absolute. So: not rhetoric *or* math, but rhetoric *before* math, which makes perfect sense if you think about how we calculate values from insurance premiums and courtroom settlements to fair wages and even the price of bread.

How does Jane Austen norm Edward's mixed feelings with respect to background actors and the realities constructed around him, and Elinor for that matter? How would these implied human beings feel with respect to their prospects and how should they be viewed? In the company of the Dashwood women and Mrs. Jennings, Elinor's brother John Dashwood situates Edward's engagement to Lucy in terms of family loyalty: "Duty, affection, and everything [i.e., money] was disregarded. I never thought Edward so stubborn, so unfeeling before" (200). In response, Mrs. Jennings resituates Edward's engagement in terms of loyalty to his word: "he has acted like an honest man!" (201). John is astonished by this retort, but does not wish to offend anybody of good fortune. So instead John notes the *extra*ordinary character of Edward's behavior in light of his mother's sudden preference for brother Robert, a preference of the sort that John declares "every conscientious, good mother, in like circumstances, would adopt" (201). Of course by this point the reader knows that John is upwardly malleable, Edward's mother cruel, and even the society matron Mrs. Jennings interested, though her "sincere" reference point coincides with the story's end. In any case, the

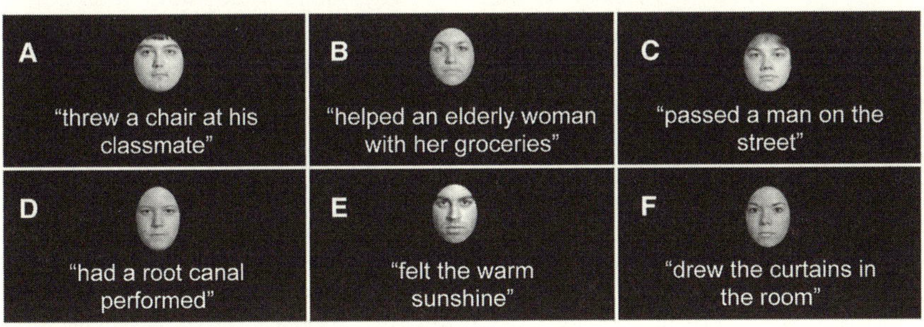

FIGURE 4.5. ". . . top-down affective information acquired through gossip influences vision, so that what we know about someone influences not only how we feel and think about them, but also whether or not we see them in the first place." (Image reproduced with permission from the American Association for the Advancement of Science)

secondary characters play a crucial role in establishing what Kahneman calls the *perceived* abnormality of an event and its affective reference points: regret, horror, outrage, and so on. And this is not to mention the even more backgrounded, but no less important, reference values lodged rhetorically beyond the passing judgment of secondary characters, in tertiary characters like Willoughby's otherwise obscure servant who appears out of nowhere to hold his horse's reins,[34] in the places like Barton Cottage that may or may not afford happiness, in the legal arrangements and inheritance guidelines that structure a novel around abnormalities and their local negotiation.[35]

In 1792 Mary Wollstonecraft famously identifies the "equivocal humiliating situation" that sets women not unlike the Dashwoods on their journey: they are dependent on the benevolence of brother John, whose marriage sets them at odds legally and therefore emotionally. It is worth quoting Wollstonecraft at some length because she provides a fitting conclusion to my argument about how fictions help establish reality.

> Girls, who have been thus weakly educated, are often cruelly left by their parents without any provision; and, of course, are dependent on, not only the reason, but the bounty of their brothers. These brothers are, to view the fairest side of the question, good sort of men, and give as a favour, what children of the same parents had an equal right to. In this equivocal humiliating situation, a docile female may remain some time, with a tolerable degree of comfort. But, when the brother marries, a probable circumstance, from being considered as the mistress of the family, she is viewed with averted looks as an intruder, an unnecessary burden on the benevolence of the master of the house, and his new partner. Who can recount the misery, which many unfortunate beings, whose minds and bodies are equally weak, suffer in such situations—unable to work and ashamed to beg? The wife, a cold-hearted, narrow-minded woman, and this is not an unfair supposition; for the present mode of education does not tend to enlarge the heart any more than the understanding, is jealous of the little kindness which her husband shows to his relations; and her sensibility not rising to humanity, she is displeased at seeing the property of *her* children lavished on an helpless sister. These are matters of fact, which have come under my eye again and again. The consequence is obvious, the wife has recourse to cunning to undermine the habitual affection, which she is afraid openly to oppose. . . .[36]

It is as if Wollstonecraft is describing the Dashwood situation exactly, but for the gratuitous comment about poor education that turns out to be inconsequential. For Austen shows in detail how such profound vulnerability also affects women who are relatively well-off in terms of such cultural capital, as we now call it. In fact the "tolerable degree of comfort" both authors attribute to some women depends upon the favor of men who might be quite good enough, that is until they marry. Exactly at that point—when the legal institution of marriage redistributes property rights to the disadvantage of a husband's immediate female relatives—sensibilities are redistributed as well, revealing the uncomfortable norm which hung in the background all along. Benevolence, which is the exclusive privilege of a house master, is undermined by his wife. Shame, which is the disproportionate burden of sisters, becomes overwhelming. Meanwhile the equalizing sensibility that might rise to humanity remains abstract and practically irrelevant, because circumstances won't allow.

Wollstonecraft is critically devastating as she, like Austen, ties sensibility directly to circumstances that appear in this case recalcitrant. At the same time these circumstances reveal themselves as negotiable, ultimately, because they appear so petty and unfair. What might first appear as the unshakable virtues and vices of humanity—benevolence and cold-heartedness—cannot be credited solely to the people in this story because they are the variables of patriarchy. But since these sentiments are so systematic, they can also be otherwise. We might even say that both Wollstonecraft and Austen are skilled negotiators as Kahneman might describe them, which is to say that they both show how reality is established with respect to its alternatives. Each author increases sympathy by showing how the victim's fate should not be considered normal, each makes it easy to imagine alternatives to the unhappy situation, and each carefully orchestrates who are focal individuals and who are background actors so that blame is removed from the suffering sisters and redistributed not to the would-be benevolent brothers, and not to the petty wives who are just looking after their offspring, but rather to the occasions that offer themselves all-too regularly. This is quite a trick, especially for a novelist whose job it is to characterize and dramatize with people, not institutions, at the focal point. Wollstonecraft can simply foreground institutional culprits like law and education. But this is only half the story as any good novelist knows. How does Austen pull it off?

We can look for example at how Austen carefully focalizes the responsibility for suffering around John Dashwood's jealous wife. No doubt Mrs. John Dashwood is indelicate when she abruptly shows up at her husband's newly inherited house as it was still occupied by the Dashwood women. But no one could dispute her "right to come" as the law speaks loudly at this moment. No doubt Mrs. John Dashwood had already registered as a potential threat to the Dashwood women. But the occasion to victimize had never presented itself up to that point. As Wollstonecraft observes, a wife's cold-heartedness and jealous behavior that undermines the habitual affection between a husband and his sisters can't be attributed to bad character primarily, but only secondarily. It is the occasion that turns a wife from vaguely unfavorable to downright destructive emotionally and otherwise. Austen puts it just this way: "Mrs. John Dashwood had never been a favorite with any of her husband's family; but she had had no opportunity, till the present, of shewing them with how little attention to the comfort of other people she could act when occasion required it" (5). That is to say the offending actions are quite squarely the result of predictable opportunities, whereas character is beside the point. So even though the reader can take some pleasure in attributing motives good and bad to a variety of characters, Austen is wonderfully cautious about distributing responsibility so that character behavior appears circumstantial, but not inevitable. Perhaps circumstances can be overcome by individuals with an extraordinary sensibility for humanity, as Wollstonecraft calls it, or what Austen considers uncommon feeling. But that is not the norm as represented, for example, by the mother in this situation. In "Mrs. Dashwood's situation, with only common feelings," unpleasantness is to be expected (5). In the hands of Wollstonecraft and Austen, virtuous sentiments appear regrettably abnormal; the norm is vicious.

Of course Wollstonecraft did not know the Dashwoods. This insufferable "situation" is shared only because it resides stubbornly in the world, not in some mishmash of personal sentiments. And Wollstonecraft was right: it is precisely such an equivocal situation that demands analysis first because it indexes how realities are established, more or less painfully.[37]

In her *Vindication*, Wollstonecraft is writing critically in a disciplinary form we would now consider social science. But our social science is not solely responsible for the analysis and criticism of social norms. Referencing Kahneman primarily next to Barsalou and Barrett, I hope to have shown in this chapter how such a situation—and the mixed feelings it engenders—deserve a certain kind of rhetorical analysis that can, indeed must, cross disciplines to do its work adequately. No single discipline has a monopoly on how norms are established referentially, and measured. At the same time we should see how rhetoric cannot be ancillary to knowledge in the disciplines, the mere linguistic form into which content is poured. In an age where the limits of rationality are newly interesting across disciplines, rhetoric is rediscovered at their very heart.

EPILOGUE

Irreconcilable Differences?

With Stephanie Preston

At least since the publication of C. P. Snow's Rede Lecture "The Two Cultures" in 1959, academics have worried about the divide between science and the humanities: the former striving to maintain clean, consistent, results by controlling the environments in which phenomena are studied, and the latter embracing natural complexities in the world, including culturally specific frames of reference that shape any particular study.[1] What has been the fate of human phenomena that—unlike particle physics or a poem respectively—do not seem to fit comfortably into one culture or the other? Emotion, for instance, is a fundamental human phenomenon that should be available to laboratory study like anything else that transpires in our material lives; at the same time, it seems defined by complexities that disappear in the controlled environment of the laboratory. The affective or "emotional turn"[2] across the disciplines has been shaped by this problem at least since the rise of neuroimaging in the 1980s, with two typical responses.

The first response to this two-cultures divide either assumes or asserts incommensurability, whereby two fields go about their business so differently that they cannot be meaningfully compared to each other, despite the fact that each might proceed rigorously according to field-specific expectations, each might produce credible data, and each might use the same key terms. An example: Neuroscience studies that examine empathy through the role of mirror-neurons or automatic neural and physiological processes[3] seem incommensurable with empathy studies in professional histories, including the work of Carolyn J. Dean (as discussed below). Where this kind of neuroscience invokes a timeframe that includes the Pleistocene approximately 10,000 years ago, and even the evolution of mammalian brains hundreds of millions of years ago, Dean ties the phenomenon of empathy to its 1873 appearance in the aesthetic/psychological vocabulary of German *Einfühlung*, and tracks its much-discussed "exhaustion" after the Holocaust.[4] Incommensurability finds a semantic-theoretical corollary in the emotion studies of Scarantino/Griffiths, who propose distinctions that would at least mitigate errors of equivocation. They argue for a clear distinction between two projects: what they call the "Folk Emotion Project" that investigates how people actually categorize emotions, can be distinguished from the "Scientific Emotion Project" designated with subscripted versions of the relevant folk emotion categories (e.g., "$anger_B$") that investigates how emotions should be categorized for scientific purposes.[5]

Deemed "consilience" by the biologist E. O. Wilson, a second influential response to the two-cultures divide has explicitly rejected incommensurability, arguing instead that we must be able to explain any human phenomenon in natural-scientific terms. Thus, according to Steven Pinker for instance, a novelistic emotion like romantic love in Jane Austen's *Sense and Sensibility* should be read in terms of evolutionary psychology that includes mating and sex differences, parent-offspring conflict, sibling rivalry, self-deception, taboo, coalitional psychology, and what he calls, following Robert Trivers, the "moral emotions," including sympathy, anger, gratitude, and guilt, which are supposed to modulate reciprocity in a manner compatible with reproductive success.[6]

Instead of proposing a third response to the two-culture problem, we offer in this essay a critique that points to its dissolution in the study of emotion. First we explain in more detail how, since the early 1980s, emotion studies have been shaped by the two-cultures alternative. A key episode in this story is a concern with the "reverse inference" problem that renders causal arguments in neuroscience suspect, and hence discourages the explicit process of inferring meaning into research studies and results, thereby sidelining a hallmark of the humanities and interpretive social sciences. Then, after reintroducing "situated cognition"—a recent approach to psychology and neuroscience that is compatible with humanities and social science research—we offer empathy as an important case study where laboratory, ethnographic, and historical work are integrated to treat socio-emotional phenomena rigorously.[7] The history of science and emotion post-1945, we finally argue, calibrates to broadly ecological concerns, and specifically, to situated models of cognition that have recently gained traction across the disciplines after a critique of the decontextualized brain.

Limits of Laboratory Science of Emotion

Where B. F. Skinner's psychology as an empirical behavioral science has been, at least since Noam Chomsky's famous 1959 review, subject to criticism for failing to study people in their natural contexts, technical constraints of neuroscience have, of late, seemed to renew the abstraction problem in psychology.[8]

1) *Technological limitations.* Subsequent to technological advances in the neuroscience of the mind that began in the early 1980s, the key tools employed by human neuroscientists to study the brain have come to include functional magnetic resonance imaging (fMRI), positron emissions tomography (PET), transcranial magnetic stimulation (TMS), and external brain recordings (e.g., EEG, ERP, MEG).[9] By and large, these techniques require subjects to sit or lie completely still during the recording, inside of or attached to machines that are cumbersome and sometimes also loud (especially fMRI, which is the most widely used). These logistical limitations make it very difficult to present subjects with naturalistic situations, scenarios, or decisions. For analytic purposes, typical stimuli must also consist of many repeated trials of events that occupy no more than a few seconds each, delivered via visual slides or audio recordings, and that allow only simple finger movements as a relevant response. Some experimenters have devised creative ways to study more natural human behavior in these environments, for example by having subjects experience the soft touch of a partner or stranger to an extended limb,[10] by showing or indicating the pain occurring to a loved one just outside of the scanner,[11] or by presenting audio clips or images of one's own baby versus an unfamiliar one.[12] But most researchers who study decision making, emotion, and prosocial behavior (i.e., empathy and altruism) have tried to solve the abstraction problem during brain scanning simply by using money as the dependent measure for value (see limitation 2, below).

Some think these types of technical limitations on neural recording will diminish as devices allow for more ecological experimental protocols. For example, near infrared spectroscopy and wireless electrodes would be less unwieldy and would permit more natural movement and interaction during the recording of neural signals. Currently, however, technologies that allow for greater degrees of movement or indirect recording also sacrifice the specificity of neural localization. For example, one can report with techniques that measure brain activity from the surface that activation was in the middle front part of the cortex rather than the lateral or posterior portion, but you can not specify that it was in BA 44. (In 1909 Korbinian Broadmann published a numbered labeling scheme for the brain that subdivided larger regions into subregions based on their cytoarchitectural properties, i.e., cell types; this nomenclature is still widely used today.) Moreover this kind of measuring device will only be able to access better a momentary ecology involving the intentional subject, but not the more durable affordances like social stratification that render the environment more legible to history: an example from chapter 2 of this book is the "happy" situation of the then-slave Equiano.

2) *Money, and the limitation of a utilitarian assumption.* Money is considered useful in this laboratory research because the decisions are considered "incentivized" (meaning that subjects are expected to give more truthful answers if their own cash rewards are at stake for doing so); moreover, the dependent measure is discrete, and easy to manipulate and measure. For example, whether or not I value an object can be indexed by how much I am willing to pay for it; how much I care for you is indexed by how much of my money I am willing to give to you; and how much I dislike something is indexed by how much I will pay to avoid it. This approach is problematic, however, because people do not actually value money linearly,[13] or the same way across individuals.[14] Even more problematic for cases like altruism and empathy, money is known to produce a mindset that specifically limits one's prosocial motives.[15] Thus, the processes involved in making a prosocial decision in the money-incentivized laboratory may not generalize to real-world situations and, therefore, should at least be augmented with ecological tasks that are performed outside of the brain scanner where the monetary decisions are particularly useful (e.g., helping someone pick up dropped papers or responding to a baby's cry).[16]

3) *Professional journal impact factor.* A researcher's ability to manage inferences is constrained by a few factors, including recent economic pressure to reach broader audiences and augment a journal's impact factor, which does not regularly distinguish between the quality and the quantity of citations to a paper. In order to reach a broad audience upon publication, or to garner media attention, researchers sometimes intentionally oversimplify while sacrificing professional knowledge about the complexity and indeterminacy of the system they study. For example, if you entitle your article something accurate but detailed like, "The temporal-parietal junction (TPJ) appears to be involved during theory of mind (ToM) processing, but maybe only because the task requires tracking movement in space over time," you are unlikely to inspire professional journal or *New York Times* editors. Instead, you can entitle your article something more exciting and forceful, which the researcher knows is also less accurate, such as, "Brain region found for the unique human ability to read minds." The exact same functional neuroimaging data

can be marshaled either way, but the more appealing claim can practically obscure the claim that is more accurate.

4) *Mass media attention.* Increasingly, decisions about how much media attention a study will attract back-propagate to influence the decisions of academic editors, granting agencies, and the researchers themselves. No doubt research grants are more often bestowed upon experienced and capable researchers of basic and applied science, but the ability to quickly publish the results in a journal or to garner supplementary personal funding (e.g., from speaking events and book contracts) is more dependent upon the ability to attract wide public attention, facilitated by simplicity and generalization that easily explains key problems of everyday experience such as how we feel, and how we act toward others. These biases seem particularly pronounced in the human social and decision sciences, which pique the interest of lay audiences, and may be responsible for increasing concerns about the validity and quality of research being published in these fields—evidenced by at least three recent data-fabrication scandals in those fields, particularly implicating researchers who are widely cited for their exciting but oversimplified claims.[17]

5) *Localization.* Lay readers are provided oversimplified explanations of the brain that ostensibly map small regions or circuits to very specific, disciplinary tasks. With terms like "the social brain,"[18] "empathy circuit,"[19] and "self-related regions,"[20] neuroscientists admittedly simplify the inner-workings of the brain and imply a straightforward mapping of tasks and functions to particular regions of the brain. However, such simplification is misleading because each brain region is adapted for particular functions—such as tracking visual motion in space—which is presumably mobilized for any task that requires that function including motor acts, person perception, and imagining the deceptions of another.[21] Even single cells as early as the visual cortex—a place where neural processing is thought to be highly structured and dependent upon the type of stimulus presented—show changes in the types of information that they process depending upon the perceptual context.[22] Thus, while any one task can show activation of a specific region in the brain when compared to another task that is construed as the "control" task, the precise level and pattern of activity will vary widely for similar processes depending on the nature of the task, the type of stimuli and decision-process used, and the nature of the control task. Moreover, because that same neural region can be activated any time the related information process is invoked, it can be engaged by a wide variety of tasks, across diverse fields and phenomena. This complexity makes it all but irresponsible to ever infer that a cell or neural region is "for" a specific psychological task, which is virtually never the case.

Beyond the appeal of manageable laboratory procedures, what is the practical appeal of brain localization in this same historical period of roughly the past three decades? Here is one historico-political hypothesis. When for instance empathy, racial hatred, or a sense of self can be located in a particular brain region, then realistically or not, prospects for intervention are in some ways easier to imagine whether that means a therapeutic cure or, alternatively, resignation in the face of biology. At the same time, it can become harder to imagine social explanations and institutional responses that seem more difficult to implement and attributable to social welfare frameworks that lost much of their appeal in the Anglo-American 1980s.[23]

6) *The limits of professional expertise.* While any one neural region can be involved in a wide variety of disciplinary psychological tasks, researchers themselves typically read and follow the research only within their own field. This exacerbates the inferential problem of associating regions with tasks, because researchers are often unaware that their "pet" brain region is also specifically implicated in relatively unrelated tasks studied in other fields, which in turn discourages more accurate and nuanced inferences about the function of the region. For example, if you follow research only in social psychology, you would perhaps never realize that the angular gyrus, which social psychologists associate with one's self or body concept, is also construed as the arithmetic region in cognitive psychology.[24] Indeed, this single parcel of cortex is known variously as the arithmetic area,[25] the metaphor area,[26] the "out of body experience" area,[27] the cross-modal area for integrating visual and spatial information with the body,[28] or the "self-relevant" area, depending on the specialty and task of the researcher and the brain hemisphere being studied. Thereby cordoned off from relevant research in neighboring disciplines, any particular research project lacks a perspective that could triangulate two functions and propose a more general function for a particular brain area, taking multiple types of tasks, and multiple research fields, into account.

7) *Reverse inference limitation*s. This problem of associating a direct mapping of task to neural function has created a backlash from the more methodologically skilled cognitive neuroscientists against "reverse inference."[29] Reverse inference discourages the assumption that the cognitive process involved in a prior study that activated a particular brain area is also responsible for the activation of that region in a subsequent study. So, for example, the following inference is discouraged in the professional neuroscientific literature: since the dosolateral prefrontal cortex (DLPFC) is often associated with the ability to manipulate information and to control the direction of one's thoughts and efforts (i.e., "working memory" and "executive control," respectively), one can infer that a social psychology task that also activated the

DLPFC requires working memory and executive control. No doubt, a ban on reverse inference of this sort is understandable in principle, because weak inferences have been common in the neuroimaging literature, as demonstrated above and by Poldrack himself. But, at the same time, such a ban enforced by editors and reviewers at prestigious journals forecloses any explicit consideration of the function of a brain area and why it was activated in the task at issue. Those who vocally oppose reverse inference, and the reviewers and editors that follow them, overlook the fact that interpretations based on extensive prior data would actually be *good* sources of information, as long as the inferences were sufficiently broad and judicious and took into account relevant data from neighboring fields and unrelated tasks. Broadly sampling from the world of data on the angular gyrus, for example, would require researchers to show more creativity and accuracy when inferring the role of this region in both arithmetic and self-reference phenomena, which could in turn tell us about the relationship between the embodied cognitive brain and psychological processes. Thus, by engaging in some responsible inference about the functions of a region, a researcher might actually be able to infer the underlying processes that various tasks require, while generating causal arguments about how a higher-order phenomenon—namely a "self" in this case—is produced, thereby implicating as well the relevant humanities and social scientific material.

So what makes the reverse inference problem stick? One reason is technocratic: vocal opponents are often from the same fraction of the profession who have access to the tools and expertise required to build empirical inferences into the tasks. No one disagrees that it would be ideal to have conditions built into the task that empirically confirm the source of a task-region association. For example, in a typical Theory-of-Mind (ToM) task, the participant must infer the beliefs of another person (e.g., "Sally") on the basis of Sally's unique second-person perspective in space and time. Thus, while the participant saw the experimenter move the cookies from the jar to the cupboard, Sally was not in the room at the time and will assume that the cookies are still in the jar. This inference requires the participant to track the movement and perception of Sally independently from the cookies, and is broadly referred to as "theory of mind." However, ToM tasks do not simply require making inferences about Sally's mental state. They also logistically require mentally tracking her movements in time and space relative to the cookies. Thus, any brain activity associated with a ToM task could be due to purely mental inferential processes (what most people assume), or it could be due to the movement tracking that is required to make this inference but is not specific to ToM tasks. To avoid reverse inference, one could add a "functional localizer" to the scan, in which the participants perform an additional block of trials that require movement tracking but not mental state attribution, such as tracking the movement of a ball behind an opaque wall.[30] One can then subtract the neural activity associated with the non-social ball-tracking task from the activity in the ToM task to determine the extent to which regions activated by the ToM task (e.g., the tempero-parietal junction [TPJ] or angular gyrus) emanated from the perceptual-motor requirements of the task in a way that is not necessarily "mental." Practically, however, the methods for generating such data are more difficult to implement, require specialized skills during analysis that many social scientists do not possess, and add time to the scan itself, which can reduce the power of the key contrast and costs significantly more money (money that the researchers against reverse inference are likely to have, but others will not).

An additional problem with the tenet that all interpretations must be empirically backed by analyses is that sometimes the most interesting results from a study are those that were not anticipated, and so were not built into the design. For example, a recent study in the Preston laboratory found activity in the insula associated with decision to acquire the most valuable goods.[31] This was unexpected because the insula was hypothesized to be more active in the subjective and affective blocks when people chose or had to relinquish goods they really wanted. A review of the literature, however, confirmed that the insula is often activated during monetary cost-benefit decisions, perhaps either because the insula is part of the neural system for assessing value on the basis of internal and affective cues, or because of affective responses to the surprisingly high- or low-value items that sometimes appeared.[32] These alternative possibilities could not be parsed in the study because the blocked design did not permit analysis of individual trials (e.g., separating valuable and cheap goods)—a design choice that is simple and statistically powerful, but would not be utilized by the more cautious, sophisticated, and highly funded researchers who vocally eschew reverse or logical inference. Therefore in this case, the authors could only speculate about the few likely reasons that the insula was active in the money condition. In fact the inference could be made on the basis of extensive empirical evidence in various human studies that the insula is involved in monetary decision tasks that are not overtly emotional. The rules against making inferences about brain activity would require, however, that the author say almost nothing about this finding if they report it at all, and a second study would have to be conducted to confirm the interesting result from the first study, doubling the cost of a methodology that is already exorbitantly expensive (at the time of writing, approximately $600 per subject, with an expected sample size of twenty to thirty participants per task, making it a minimum of $12,000 per experiment and usually over $100,000 due to the ancillary costs of hiring a trained full-time staff that run and analyze imaging studies).

8) *Technocratic hierarchy.* This situation generates a sharp divide in which the few highly funded and technically proficient labs that manage multiple large federal grants and a large staff of personal secretaries, engineers, statisticians, and physicists control and generate the vast majority of published studies on the brain—these also happen to be the researchers who most often eschew the simpler techniques and logical inferences, which nonetheless can produce useable information that moves the field forward if done responsibly.

Fear Situated

We have now outlined some important factors that have intensified the brain-abstraction problem along with the rise of neuroimaging over the last three decades, exacerbating a two-cultures problem in and around brain science. During this same historical moment, however, a powerful alternative has persisted in philosophy of mind that references Wittgenstein and Vygotsky, phenomenological traditions after Martin Heidegger, and the perceptual ecology of American psychologist J. J. Gibson. Now we will consider how this approach can meet the professional expectations both of neuroscience, and of the humanities and interpretive social sciences. Case studies include first "fear" as it has been treated in the Phelps and Barsalou laboratories, then "empathy" as studied by Preston and colleagues. In both cases, the emotional topics—fear and empathy—speak to pressing post-war social concerns: racism in the first instance, and humanitarianism in the second. Hence our central historical claim: The *topics* of laboratory work on the emotional brain coincide meaningfully with particular socio-historical concerns that show up at the margins of experiment design and interpretation.[33]

In a widely cited 2000 article that appeared in the *Journal of Cognitive Neuroscience*, New York University cognitive scientist Elizabeth A. Phelps and her colleagues used fMRI to examine the neural response of "White American" subjects who viewed images of African Americans and White Americans.[34] Phelps and colleagues say that their study in social cognition was motivated by the remarkable difference between, on the one hand, a steady decline in self-reported racial prejudice, and, on the other hand, persistent "unconscious" racial prejudice directed at Black people (729). One of two primary goals in this study was to examine the neural correlates of responses to racial groups, focusing on the amygdala because it appears to be involved in emotional learning especially in relation to fear, memory, and evaluation (729). Using fMRI brain imaging technology, the Phelps group investigated correlations amongst scores on the Modern Racism Scale (commonly used to measure conscious, self-reported beliefs and attitudes toward Black Americans, 736), unconscious measures of startle response and bias, and fMRI-derived amygdala activity in White Americans responding to Black and White male faces with "neutral" expressions.

In experiment 1, the faces presented belonged to individuals who were unfamiliar to the subjects. In experiment 2, the faces belonged to "famous and positively regarded" Black and White individuals (730) including Arsenio Hall, Bill Cosby, Magic Johnson, Martin Luther King Jr., and Colin Powell on the one hand; Conan O'Brien, Tom Cruise, Larry Bird, John F. Kennedy, and Norman Schwarzkopf on the other (730, 736). According to Phelps, these two studies showed for the first time that members of Black and White social groups can "evoke differential amygdala activity" and that "this activity is related to unconscious social evaluation" (734). In experiment 1, the strength of amygdala activation to Black-versus-White faces was correlated with the two unconscious measures of race evaluation, but not with the conscious expression of racial attitudes from the self-report questionnaire. In experiment 2, these patterns were not reproduced when the faces observed belonged to familiar and positively regarded Black and White individuals (734).

Interestingly, Phelps et al. are cautious about making socio-historical inferences connecting the indirect measures and the brain data, while nonetheless placing the data within a larger context implicated by the provocative topic itself. They qualify: "these data cannot speak to the issues of causality" (734), but go on to posit that "both amygdala activation as well as behavioral responses of race bias are reflections of social learning within a specific culture at a particular moment in the *history of relations between social groups*. . . . Unless one is socially isolated, it is not possible to avoid acquiring evaluations of social groups, just as it is not possible to avoid learning other types of general world knowledge" (734, our italics). So in one influential study we get a two-cultures impasse in dramatic form.

On the one hand it is suggested that the history of relations between Blacks and Whites should count as a significant causal factor producing unconscious racial bias in a particular subject, as measured in amygdala response correlated with certain unconscious measures. On the other hand, Phelps and her colleagues insist that their data *cannot* speak to issues of causality. Productively this study comes right up to the brink of a two-cultures divide, then falters. What causes racialized fear? Phelps et al. gesture vaguely toward history, but give us concretely only neuroscientific confirmation that such racialized fear exists (despite a subject's explicit protests to the contrary). The problem with this disconnect between vague history and concrete neuroscientific data is that the protocol for connecting the two—let us call it a mediating field of socio-historical phenomenon—is not sufficiently articulated to shape the laboratory work and its interpretation.

By drawing on critical race theory and rhetorical studies, we might say that the White, avowedly pro-Black subject who startles more quickly, and negatively associates more readily when presented with an unfamiliar Black face, belongs to a broader social phenomenon that must be analyzed at the historical level that exceeds any particular subject. The Phelps study gives us a subject who demonstrates particular responses coded as "racist," but to understand or even to initially identify those responses, the phenomenon (racism) must have reference points beyond the brain physiology ultimately studied. Racism is not located primarily in the brain. It is located primarily in the racist institutions—like US race-based slavery and subsequent Jim Crow laws—that exceed any particular subject and can persist without any particular kind of incorporation or brain state. In this case, both sides of the equation are necessary for the study to matter. Without the physiology, no doubt, claims of declining racism might be overstated owing to the ideological prestige of anti-racist discourse. And without studies broadly conceived in the humanities, such physiological evidence could never show up because we would not know what we were looking for, or at least we wouldn't care. A more robust brain science of emotion—in this case racialized fear—would take seriously in design and interpretation the "history of race relations" invoked incidentally by Phelps.

Scholars in the humanities and interpretive social sciences might also be interested in how televisual and other sorts of mediation show up in experiment 2, where faces belonging to "familiar and positively regarded" Black and White individuals significantly mitigate the racism identified in experiment 1. But instead of emphasizing like Phelps et al. the anti-racist implications of this finding, which would suggest that familiarity mitigates racism, such scholars might underscore the socially mediated quality of each experimental stage. This would include category selection ("Black" and "White," which are socio-historical categories not natural kinds), image selection ("Black" and "White" image attribution, which will vary significantly according to the cultural situation of the laboratory and its experiment and might look different in Brazil, for instance), and also the very concept of "familiarity," which in this case has nothing to do with face-to-face encounters, personal acquaintance, or family or communal relationships, and everything to do with mediation as *media* that locates and dis-locates images according to complex regimes of publicity and capital of the sort studied thoroughly by media and rhetoric scholars. Hence one might question, for instance, the Phelps presentation of stimuli that equates a list of Black and White individuals whose faces are supposed to portray rough equivalents in degrees of fame, age, and achievement (733). Only a broader perspective can foreground the fact that, although Martin Luther King Jr. and John F. Kennedy might provide facial images roughly equivalent within strict parameters defined by the experiment, fame and achievement and even longevity are unevenly distributed across Black and White populations, which means at least that a famous and familiar Black face carried with it into the laboratory certain kinds of markedness, whereas the supposedly equivalent White face did not. To put this another way, Phelps may be correlating *celebrity* with unconscious racial attitudes, not face-to-face familiarity. Her study does not adequately specify "familiarity" so no meaningful analytic distinction can be drawn between the familiarity of an Arsenio Hall face seen by a white subject only in the context of television and accompanying media, versus the familiarity, say, of Equiano to his slave master Henry Pascal. In the context of racialized slave culture for instance, "familiarity" did not mitigate racism, which suggests that the term must be deployed with some more historical rigor if it is to be socially useful in this way suggested by Phelps. More broadly the social cognition study of racialized fear must be "situated" with greater rhetorical specificity if it is to produce results that speak to causality. When it comes to emotion studies, it would appear, both sides of the two-culture divide suffer the shortcomings of a shared phenomenon not treated as such.

Referencing Wittgenstein on natural language, Lawrence Barsalou and colleagues also study fear-as-concept, in the effort to critique and dissolve aspects of the two-culture problem that face Phelps and others. In a summative essay, "Grounding Emotion in Situated Conceptualization" (2011), Wilson-Mendenhall, Barsalou, and colleagues critique the "basic emotion" approach made famous by Paul Ekman, which assumes that "emotions reflect inborn instinct, and that the mere presence of relevant external conditions triggers evolved brain mechanisms in a stereotyped and obligatory way (e.g., a snake triggers the fear circuit)" (1105). Barsalou and colleagues instead assume that emotions are not fixed, discrete categories, but are rather concepts that refer to entire situations, and thereby represent "settings, agents, objects, actions, events, interoceptions, and mentalizing" (1107). Thus the *fear* of a runner who becomes lost on a wooden trail at dusk, for instance, differs from the *fear* of someone unprepared to give an important presentation at work. In this latter situated conceptualization, they explain, "a different set of concepts represents the situation, including *presentation*, *speaking*, *audience*, *supervisor*, and many others" (1108). In essence, their laboratory experiment explores this hypothesis that the basic emotion approach is wrong since, in fact, a situated conceptualization "produces the emotion" (1110). Or to put this another way, the situation in which an emotion concept is experienced shapes how the emotion is instantiated in the brain (1120). For example, the observation that social fear activated clusters in the left dorsolateral prefrontal cortex and inferior frontal gyrus suggests that social fear, as opposed to physical fear, requires more executive control to cope with threatening social evaluations (1124).

The situated emotion approach of Barsalou, Barrett,[35] Campos,[36] and others, does challenge the two-cultures problem by explicitly incorporating language and audience considerations into experimental design, and into interpretation. However, Barsalou's modular, functionalizing focus on situated conceptualization is forced to rely upon elliptical figures ("and so forth," "and many others") to articulate a background that recedes indefinitely into what we would consider *non*-conceptual affordances located in the environment, including the social environment like Phelps's "history of race relations." Racialized fear, for example, is not just a concept that refers to an entire situation and thereby represents "settings, agents, objects, actions, events, interoceptions, and mentalizing." Racialized fear, our next theorists would argue, is actually built into a social situation that can affect people directly, without requiring a persistent or pervasive conceptual representation.

Instead of working exclusively in the laboratory, Paul Griffiths and Andrea Scarantino track "emotions in the wild" as they try to shift theoretical focus to neglected phenomena such as anger in a marital quarrel or embarrassment while delivering a song to an audience (438). Emotional content, they argue, has a "fundamentally pragmatic dimension, in the sense that environment is represented in terms of what it affords the emoter in the way of skillful engagement with it" (441). In an effort to theorize cognition beyond the brain, Griffiths and Scarantino consider the "active contribution of the environment," such as confessionals in churches that enable certain kinds of emotional performance that they call synchronic scaffolding after Vygotsky, and the broader Catholic culture that supports the development of the ability to engage in the emotions of confession, which they call diachronic scaffolding (443). They also consider material factors, including "emotional capital," such as the emotional resources associated with having a specific social status, gender, disability, etc. (444). But what would the science of emotion look like if we genuinely tried to integrate laboratory, ethnographic, and historical considerations?[37] To address this question, we will conclude with an example from the Preston Ecological Neuroscience Lab that specializes in studying how empathy is instantiated in the body and brain. In doing so we try to learn from, and advance, the emotion research of Phelps, Barrett, Barsalou, Griffiths/Scarantino, and others toward the dissolution of a two-cultures divide.

Empathy Situated

What is empathy? Mirror-neuron research casts empathy as a non-cognitive and trans-primate phenomenon whereby observing an action (e.g., cup raising) fires a significantly similar set of neurons in the observer as when they perform that same action. But isn't a rigorous treatment of the phenomenon "empathy" obligated to explain at least why certain actions elicit empathic responses while others do not? Otherwise one risks a circular definition where empathy is everything that involves mirror-neuron activity, and all mirror-neuron activity is by definition empathic. To cite an obvious example, why does empathy appear in some suffering circumstances but not others, even when the objective conditions appear the same? At this point case studies become crucial.

A series of studies by Preston, Hofelich, and colleagues involves videos of real hospital patients who are discussing their experience with serious or terminal illness.[38] While all of these patients are hospitalized for very serious medical conditions, they vary widely in terms of the emotions they invoke during the interview, from distraught crying throughout, to laconic resistance to talk, to resilient sadness potentiated by affiliation. But it is not only the patients who differ in terms of their emotional profile; observer reactions also differ significantly. Some patient-observer combinations produce empathy and the desire to help, others do not. The variance across observers is particularly pronounced towards the most distraught patients on one side of the spectrum, and the most reticent patients on the other. Preston and her colleagues found that most observers in the study empathize with the former and not the latter. But observers who are currently experiencing their own intense negative affect due to depression report less sympathy for the distraught patients—understood as empathy plus the desire to help—while those who report that they often "take another's perspective" in daily life report greater sympathy for the reticent patient who is less demonstrative but does appear to have deep emotions and need.

Why so? If empathy were a non-cognitive, automatic, and context-insensitive phenomenon, wouldn't pain trigger pain, sadness trigger sadness, and in a linear fashion that could make both empathy reports and sympathetic behaviors more predictable over human subjects and situations? How does one explain, for instance, the "reticence" factor that correlates in a complicated fashion with apathy and empathy?

At this point we consider situational factors that include a high degree of historical specificity. Carolyn J. Dean carefully tracks how "reticence" became an important quality in the recent production of empathy, or compassion, in one crucial post-war emotional domain: Holocaust victimization.[39] Her key point relevant to the Preston study is that—as a result of false victim controversies and the surrounding melodrama—reticence has become a key marker of authentic suffering and hence a minimum requirement for empathic or compassionate response in the case of survivor stories. However, Dean is careful to point out that, in this case and others, a situational formula is not forthcoming if we want to treat an emotional phenomenon accurately. Instead a range of humanistic and social scientific tools must be brought to bear if we want to understand with some accuracy how empathy does or does not function in any particular case. "The preference for minimalist style in victim testimonies," Dean reminds us, "is but one component of a much broader discourse about the exemplary victim. Over and over, the presumptive cultural demands of false victims (here defined by the dramatic contrast between them and the putatively 'extraordinary reticence' shown by many survivors) trump any inquiry into how some victims are deemed more credible than others since we are presumed to know that 'real' survivors would by definition be reluctant to seek attention. Reticence thus distinguishes real and false victims" (396).

So the larger point is not that the emotional reticence of Preston's hospital patients and the emotional reticence of Dean's survivors are equivalent. At least one significant difference is that across the Preston and Hofelich studies, the distraught patients actually receive on average the most sympathy and help—though they can also elicit strong negative responses from minorities of the subjects. Presumably, their high and overt negative emotion is necessary to know that they are in need, and deemed permissible since they are currently experiencing the distress of illness or pending death. However, the resilient patients who exhibit little distress and much more positive emotion reliably receive the second most sympathy and the reticent the least—in contrast to a situation-insensitive supposition that reticence necessarily fosters sympathy.

More important, for us, is the general injunction to consider the relevant emotional genres and styles, and the specific injunction to consider how a research question about empathy is located in a particular historical and cultural situation where we cannot take for granted the transparency of our key term. For instance in the hospital study, it turns out that gender expectations play a key role in differentiating the extent to which people want to help the distraught versus the resilient patients, even though all observers agree that the former need more help. This hospital study thus cannot be gender-neutral without distorting the apathy-empathy phenomenon of interest in the first place. Meanwhile in the reception of a survivor's tale, anxiety of transmission now becomes prominent as historians and their audiences worry about losing the subjects of study. In both cases, but in different ways, any neuroscience of empathy would have to take these historical considerations seriously at the level of both experimental design, and interpretation. At the same time, a humanities perspective on the shared key term "reticence" invites a line of inquiry that does not assume the link to be accidental, but instead tracks the interference of suffering situations across particular case studies. Finally, with Dean, we can point toward a humanities construction of dignity that lays out broadly how victims—of systematic oppression or "natural" circumstances beyond one's control—are constituted, along with the agencies that are supposed to respond.[40] At the very least, Preston's study suggests how it is not necessarily impractical to consider history and situation when studying empathy scientifically, and the hope is that such considerations enhance the quality of the conclusions.

Summary and Conclusion

We began this essay with a description of the two-cultures divide, in which academics of natural science on one side, and the humanities and interpretive social sciences on the other, appear to be at an impasse *explained* by incommensurability, but *addressed* most obviously by consilience in the reductive tradition of E. O. Wilson. However, experiments—even within the context of neuroscience, which is focused on the underlying mechanisms for behavior—can proceed more ecologically and thereby provide new insight into brain-body-world dynamics, especially as they are implicated in important social phenomena.

Significant barriers remain. Scientists must come up with experimental protocols that are ecological, situated, and statistically powerful, within the confines of existing technology. At the same time, those of us in the humanities must avoid the all-or-none proposition that their topics—for example, racial hatred, empathy, or the self—must be completely available to experimental protocol, or completely unavailable. Because of the nature of the experiment, which holds most things constant in order to be able to make causal claims about the role of the key variable of interest, one usually cannot study many, possibly competing and interacting variables at the same time.

Thus, for example, to address our critique of Phelps and colleagues would take dozens of studies instead of just the one, which was already suitable for publication in the flagship journal for the cognitive neurosciences.[41]

We hope that this essay, written by a rhetorician and a neuroscientist, can serve as an example of two-cultures collaboration at its early stages. In situated cognition we identify an approach that invites such collaboration. Given scientific concerns, it would be misleading to study "cognition" without considering work in neuroscience. At the same time science cannot have the last word without incorporating some of the situational concerns of the humanities, because "situations" are environmental, including a social environment that can be quite specific like empathy exhaustion, and temporal, which sometimes means historical in the precise, professional sense of the word.[42] Ultimately the two-cultures divide dissolves when shared phenomena like human emotions are treated with the methodological diversity and cross-disciplinary conversation their complexity deserves.

Notes

Introduction

1. C. D. Wilson-Mendenhall, L. F. Barrett, W. K. Simmons, and L. W. Barsalou, "Grounding Emotion in Situated Conceptualization," *Neuropsychologia* 49, no. 5 (2011): 1105–1127, at 1110.

2. Jane Austen, James Kinsley, and Deirdre Shauna Lynch, *Persuasion* (Oxford: Oxford University Press, 2004), 198.

3. Lauren Berlant, *Cruel Optimism* (Durham: Duke University Press, 2011), 5.

4. Along with Chandler (cited next), some key relevant works on sentimental literature include Graham J. Barker-Benfield, *The Culture of Sensibility: Sex and Society in Eighteenth-Century Britain* (Chicago: University of Chicago Press, 1996); Markman Ellis, *The Politics of Sensibility: Race, Gender and Commerce in the Sentimental Novel* (Cambridge: Cambridge University Press, 2004); and Lynn Festa, *Sentimental Figures of Empire in Eighteenth-Century Britain and France* (Baltimore: Johns Hopkins University Press, 2006).

5. James Chandler, *An Archaeology of Sympathy: The Sentimental Mode in Literature and Cinema* (Chicago: University of Chicago Press, 2013).

6. In *Blake's Agitation: Criticism and the Emotions* (Baltimore: Johns Hopkins University Press, 2013), 12–13, Steven Goldsmith cites Chandler as he describes this microscopic process with regard to Blake.

> Exposing the scaffolding beneath its illusion of spontaneity, Blake allows us to glimpse that sentiment is "made," and he can do this only by "unmaking" it from within, by exploiting the loose joints of its grammatical construction. Tracking this interior process involves Chandler in a microscopic analysis of representative lyrics from *Songs of Innocence and of Experience*, for the kinds of abrasions that "denaturalize sentiments" turn on the disjunctions of single words or phrases. Consider, for instance, the word "tender" in a line from "The Shepherd": "And he hears the ewes tender reply." Without the expected apostrophe in "ewes," as Chandler notes shrewdly, "'tender' becomes an economic verb rather than a sentimental adjective." By inviting this small grammatical ambiguity, Blake not only emphasizes "the madeness of the sentiments" but also directs our attention to sentimentalism's underlying economic function—to smooth the social transactions necessary to make commerce seem natural.
>
> — Citing James Chandler, "Blake and the Syntax of Sentiment: An Essay on 'Blaking'

Understanding," in *Blake, Nation and Empire*, ed. Steve Clark and David Worrall (Basingstoke: Palgrave Macmillan, 2006).

7. Paul Griffiths and Andrea Scarantino, "Emotions in the Wild," in *The Cambridge Handbook of Situated Cognition*, ed. Philip Robbins and Murat Aydede (Cambridge: Cambridge University Press, 2009).

8. Clifford Geertz, *The Interpretation of Cultures: Selected Essays* (New York: Basic, 1973), 30.

9. Sarah Fielding and Peter Sabor, *The Adventures of David Simple: Containing an Account of His Travels Through the Cities of London and Westminster, in the Search of a Real Friend; And, the Adventures of David Simple, Volume the Last: in Which His History Is Concluded* (Lexington: University Press of Kentucky, 1998).

10. See for instance Steven Pinker, "Toward a Consilient Study of Literature," in *Philosophy and Literature* 31, no. 1 (2007): 161–177; reviewing Jonathan Gottschall and David Sloan Wilson, eds., *The Literary Animal: Evolution and the Nature of Narrative* (Evanston, IL: Northwestern University Press, 2005).

11. Lawrence Stone, *The Family, Sex and Marriage in England 1500–1800* (New York: Harper & Rowe, 1977), 282.

12. See for instance the book review by Alan Macfarlane in *History and Theory: Studies in the Philosophy of History* 18, no. 1 (1979): 103–126.

13. William M. Reddy, *The Making of Romantic Love: Longing and Sexuality in Europe, South Asia & Japan, 900–1200 CE* (Chicago: University of Chicago Press, 2012). Eighteenth-century European sentimental literature, notably Samuel Richardson's *Pamela*, marks for Reddy "a rethinking of the relationship between love and social rank" (380).

14. Laurence Sterne, *A Sentimental Journey through France and Italy by Mr. Yorick* with *The Journal to Eliza* and *A Political Romance*, ed. with an introduction and notes by Ian Jack (Oxford: Oxford University Press, 1968), 18–19. Compare Walter Shandy's polemical point in the Toby-Wadman affair: "love, you see, is not so much a SENTIMENT as a SITUATION, into which a man enters, as my brother Toby would do, into a corps—no matter whether he loves the service or no—being once in it—he acts as if he did; and takes every step to shew himself a man of prowesse." Laurence Sterne and Ian C. Ross, *The Life and Opinions of Tristram Shandy, Gentleman* (Oxford: Oxford University Press, 2009), 475.

15. Perhaps like Thomas Reid, whose *An Inquiry Into the Human Mind on the Principles of Common Sense* was just published in 1764 and has plenty to say about the intervention of judgment upon that which is perceived directly. Reid's "direct perception," incidentally, will have some influence upon the perceptual psychologist J. J. Gibson, who starts to appear prominently in chapter 3 of my project.

16. Sianne Ngai, *Ugly Feelings* (Cambridge, MA: Harvard University Press, 2005), 81–82.

17. Amongst other places "prosthetic emotion" is discussed in Susan Miller, *Trust in Texts: A Different History of Rhetoric* (Carbondale: Southern Illinois University Press, 2008), 14, referencing Cynthia Haynes, "pathos@play.prosthetic.emotion," in *Works and Days* 13, no. 1–2 (1995): 261–276. I am also indebted to Libby Catchings's dissertation work on this topic.

18. Philip Fisher, *The Vehement Passions* (Princeton: Princeton University Press, 2002), 142.

19. See especially Robert B. Zajonc's influential article, "Feeling and Thinking: Preferences Need No Inferences," *American Psychologist* 35, no. 2 (1980): 151–175.

20. Ruth Leys, "The Turn to Affect: A Critique," *Critical Inquiry* 37, no. 3 (2011): 434–472, at 437.

21. Brian Massumi, "The Autonomy of Affect," *Cultural Critique* 31 (1995): 83–109. See

also *Parables for the Virtual: Movement, Affect, Sensation* (Durham: Duke University Press, 2002).

22. Lauren Berlant, "Nearly Utopian, Nearly Normal: Post-Fordist Affect in *La Promesse* and *Rosetta*," *Public Culture* 19, no. 2 (2007): 273–301.

23. Raymond Williams, *The Long Revolution* (Ontario: Broadview Press, 2001 [1961]), 321.

24. Jordanna Bailkin highlights how decolonization and the expansion of the postwar welfare state inaugurated new emotional ties that sought to bind together young elites in Britain with those in the former colonies, thus creating new types of social bonds now understood within an evolving international human rights framework. She shows how the rise of the welfare state in some sense transformed all Britons for the first time into both givers and receivers of aid, and how a postwar regime of democratized and universal well-being was in part established on the uneven terrain of decolonization. See Frank Biess and Daniel M. Gross, eds., *Science and Emotion after 1945: A Transatlantic Perspective* (Chicago: University of Chicago Press, 2014), 278–299.

25. There are some exceptions to this mystification—for example, Fredric Jameson on a Paris fish market in Zola. *The Antinomies of Realism* (London: Verso, 2013), 53–54. But this convincing read comes after Jameson offers a systematically contrasting table in the spirit of Massumi: EMOTION on one side as system, nomenclature, representation, etc., and AFFECT on the other side as chromaticism, bodily sensation, sense-data, etc. (44).

26. Kent Puckett, *Bad Form: Social Mistakes in the Nineteenth-Century Novel* (Oxford: Oxford University Press, 2008), 11, 53–59.

27. Especially David Chalmers, forward to Andy Clark, *Supersizing the Mind: Embodiment, Action, and Cognitive Extension* (Oxford: Oxford University Press, 2008); Michael Wheeler, *Reconstructing the Cognitive World: The Next Step* (Cambridge, MA: MIT Press, 2005); Evan Thompson, *Mind in Life: Biology, Phenomenology, and the Sciences of Mind* (Cambridge, MA: Belknap, 2007).

28. Jakob von Uexküll, *A Foray into the Worlds of Animals and Humans*, trans. Joseph D. O'Neil, introduction by Dorion Sagan (Minneapolis: University of Minnesota Press, 2010 [1934]), 172.

29. Michel Serres, *The Natural Contract*, trans. Elizabeth MacArthur and William Paulson (Ann Arbor: University of Michigan Press, 1995), 33, offers a different critique: "forget the word *environment*" (emphasis author's). "It assumes," he writes, "that we humans are at the center of a system of nature. This idea recalls a bygone era, when the Earth . . . reflected our narcissism, the humanism that makes of us the exact midpoint or excellent culmination of all things." Quoted in object-oriented literary study: Julian Yates, *Error, Misuse, Failure: Object Lessons from the English Renaissance* (Minneapolis: University of Minnesota Press, 2003), 1. The literature on object-oriented philosophy à la Serres, Latour, Harman, or Bennett grows quickly. In rhetorical studies per se, see especially Thomas Rickert, *Ambient Rhetoric: The Attunement of Rhetorical Being* (Pittsburgh: University of Pittsburgh Press, 2013).

30. Sigmund Freud, *Das Unbehagen in der Kultur* ("The Uneasiness in Culture") translated traditionally as *Civilization and Its Discontents* via James Strachey, *The Standard Edition of the Complete Psychological Works of Sigmund Freud*, Vol. 21, 1927–1931 (New York: Norton, 1961): "With every tool man is perfecting his own organs, whether motor or sensory, or is removing the limits to their functioning. . . . Writing was in its origin the voice of an absent person; and the dwelling-house was a substitute for the mother's womb, the first lodging, for which in all likelihood man still longs, and in which he was safe and felt at ease" (43). David Simpson, in his

book *Situatedness: or, Why We Keep Saying Where We're Coming From* (Durham: Duke University Press, 2002), provides a genealogy for this popular phenomenon as it runs through social science disciplines of the Enlightenment; his book is about "how people have tried and still try to set limits, to make claims, and to find comfort" (4).

31. But in "The Situation as a Function of Human Needs" from *What Computers Still Can't Do: A Critique of Artificial Reason* (Cambridge, MA: MIT Press, 1992 [1972]), 272–280, Dreyfus argues that human beings do not "arbitrarily adopt values which are imposed by their environment" (277). His first example is a man who falls in love with a particular woman and thus becomes the sort of person that needs that specific relationship and must view himself as having lacked and needed this relationship all along. "In such a creative discovery," explains Dreyfus, "the world reveals a new order of significance which is neither simply discovered nor arbitrarily chosen" (277). Here Dreyfus references Søren Kierkegaard on the leap of existence where one's personality or self is redefined and "everything in a person's world gets a new level of meaning" (277). Indeed such a change, by modifying a person's concerns, "changes the whole field of interest in terms of which everything gets its significance" (277). Dreyfus also references Thomas Kuhn on revolutions at the conceptual level, and he might well have referenced Merleau-Ponty on political revolution as I will in chapter 2.

32. Franco Moretti, *The Bourgeois: Between History and Literature* (New York: Verso, 2013), 45.

33. In fact Ahmed provides a beautiful study of philosophy and the concealment of domestic labor as she tells the story of phenomenologist Edmund Husserl and his writing desk supported by his wife in the other room preparing supper and the like. See *Queer Phenomenology: Orientations, Objects, Others* (Durham: Duke University Press, 2006), 28–30.

34. Recently Jonathan Kramnick has brought our attention to Locke's feeling of "uneasiness" which is a response to circumstances and counts in the order of causes that includes the mind "along with other parts of the world." See Jonathan Kramnick, "Uneasiness, or Locke among Others," in *Actions and Objects from Hobbes to Richardson* (Stanford: Stanford University Press, 2010), especially 158. As I am particularly interested in the parts of the world where social institutions distribute uneasiness along lines of power, I would instead highlight Locke on "fantastical uneasiness": "The ordinary necessities of our lives, fill a great part of them with the *uneasiness of Hunger, Thirst, Heat, Cold, Weariness* with labor, and *Sleepiness* in their constant returns, etc. To which, if besides accidental harms, we add the fantastical *uneasiness* (as it's after *Honour, Power,* or *Riches* etc.) which acquir'd habits by Fashion, Example, and Education have setled in us, and a thousand other irregular desires, which custom has made natural to us, we shall find, that a very little part of our life is so vacant from these *uneasinesses,* as to leave us free to the attraction of the remoter absent good." John Locke, *An Essay concerning Human Understanding* (Oxford: Oxford University Press, 1975), XXI.45; 261.

35. Note how I will move between emotion and feeling based primarily on common expression and fixed phrases, while some of the technical and historical arguments distinguishing emotion, feeling, affect, and passion appear in the footnotes only when necessary.

36. Following a famous 1968 article by Lloyd Bitzer, the "rhetorical situation" includes an *exigence* (in terms of my project a basic discomfort), an *audience,* and a set of *constraints* that are part of the situation because they have the power to constrain decisions and actions needed to modify the exigence: Bitzer lists beliefs, attitudes, documents, facts, traditions, images, interests, motives, and the like (8). Bitzer's definition of the exigence is particularly relevant: "Any exigence is an imperfection marked by urgency; it is a defect, an obstacle, something waiting to be done, a thing which is other than it should be. In almost any sort of context, there will be numerous

exigences, but not all are elements of a rhetorical situation—not all are rhetorical exigences." Lloyd F. Bitzer, "The Rhetorical Situation," *Philosophy and Rhetoric* 1, no. 1 (1968): 1–14, at 6.

37. Next to Burke, I'm thus invoking Martin Heidegger and especially his 1924 lectures on Aristotle's *Rhetoric*, where rhetoric is grounded in slightly more paranoid terms: lack of composure [*Aus-der-Fassung*] brought on by terror forces us to decide who is with us and who is against us, and then to articulate our concerns. In fact the reasoned outcomes of judicial and deliberative rhetoric are grounded in more basic rhetorical commitments shaping community (*Mitsein*), such as the cleavage of friend and enemy. See Martin Heidegger, *Grundbegriffe der aristotelischen Philosophie: Marburger Vorlesung Sommer Semester 1924*, Gesamtausgabe Vol. 18, ed. Mark Michalski (Frankfurt a. M.: Klostermann, 2002), 261.

38. Kenneth Burke, *The Philosophy of Literary Form: Studies in Symbolic Action* (Baton Rouge: Louisiana State University Press, 1973 [1941]), 60–61.

39. This analysis is indebted to Jami Bartlett, who has recently written the book on nonreferring terms, *Object Lessons: The Novel as a Theory of Reference* (Chicago: University of Chicago Press, 2016).

40. Hence I offer an alternative to Chandler's work on mixed feelings in the line of Schiller, whose treatise *On Naïve and Sentimental Poetry* (1795) focuses on the reflexivity of an emotional experience as when Tristram expresses remorse for pleasantries around poor Maria of Moulines. Chandler calls this a "performed ambivalence" (157), or after Schiller a mixture of feelings (*zwei streitenden Empfindungen*, 153) that derives from the reflexivity of the generative process (158). My analysis also notices the reflexivity of mixed feelings as they compose character and draw our attention to the phenomenon, but I argue that the components of this mixture derived from conflicts that reside outside of the character, in a world that imposes a range of emotional alternatives appearing one way or another. In fact Chandler, following Ian Hacking, comes close to this observation in his later discussion of sentimental probability as it plays out in the *Sentimental Journey* scene where La Fleur's three degrees of exclamation are analyzed: *Diable*, *Peste*, then probably *Nom de Dieu*. "Fortune or misfortune acquires a kind of measure in the marked levels of expressive exchange between agents. The assumption that 'accidents' move us is elaborated by way of the conceit that in some 'casts' of the dice Fortune 'provokes' or calls us forth. . . . Emotion can be thus closely keyed to degrees of improbability" (227).

41. Daniel Kahneman and Dale Miller, "Norm Theory: Comparing Reality to Its Alternatives," *Psychological Review* 93, no. 2 (1986): 136–153.

42. This is a piece that expands work we did together for the Frank Biess and Daniel M. Gross edited volume *Science and Emotion after 1945: A Transatlantic Perspective* (Chicago: University of Chicago Press, 2014).

Chapter One

1. Charles Darwin, *The Expression of the Emotions in Man and Animals* (Oxford: Oxford University Press, 1998 [1872]), 139–140; hereafter abbreviated *E*. For reasons that will become obvious, I cite the third edition throughout this chapter.

2. Alva Noë, *Out of Our Heads: Why You Are Not Your Brain, and Other Lessons from the Biology of Consciousness* (New York: Hill and Wang, 2009).

3. Ian Hacking's 1998 *Times Literary Supplement* review is prescient; among other things he notes how efforts in psychology to eliminate emotion as a natural class recalls the young logical positivist A. J. Ayer eliminating metaphysics. Before the location of emotion in recent advances in brain-imaging techniques, it seemed faculties of our rational animal (including

knowledge, perception, judgment) held out more promise for scientific study. See Ian Hacking, "By What Links Are the Organs Excited?" *Times Literary Supplement* no. 4972 (1998): 11–12. For an extended treatment of emotion and postwar rationalism see Frank Biess and Daniel M. Gross, eds., *Science and Emotions after 1945: A Transatlantic Perspective* (Chicago: University of Chicago Press, 2014).

4. See Paul Ekman, "How to Spot a Terrorist on the Fly," *Washington Post*, October 29, 2006, www.washingtonpost.com/wp-dyn/content/article/2006/10/27/AR2006102701478.html.

5. For Ekman's blog on the show, see www.paulekman.com.

6. The outstanding critical work on this topic is Sander L. Gilman, *Seeing the Insane* (Lincoln: University of Nebraska Press, 1996 [1982]). Among other things, Gilman tracks, through Darwin's correspondence with the psychiatric physician James Crichton Browne, Darwin's increasing skepticism about the utility of photographs expected to deliver stable referents for the scientific study of emotion, especially insofar as they might provide evidence for the insane and idiotic as the missing link to our emotional past.

7. For a post-Darwinian history of faces in the psychology of emotion, see John McClain Watson, "From Interpretation to Identification: A History of Facial Images in the Sciences of Emotion," *History of the Human Sciences* 17, no. 1 (February 2004): 29–51.

8. For detailed treatment of Darwin's groundbreaking use of photography in the *Expression*, see Phillip Prodger, *Darwin's Camera: Art and Photography in the Theory of Evolution* (Oxford: Oxford University Press, 2009). Prodger underscores Darwin's "rhetorical abilities" (30), which extended to his incorporation of photographs in the *Expression*. Though like Ekman we tend to read the history of photography retroactively, Prodger underscores how evidence and illustration are blurred in the *Expression* because there was essentially no precedent for the acceptance of photography as scientific data. Like neighboring printed illustrations that recalled earlier physiognomic and expressivist traditions (for example, Lavater and Le Brun respectively), photographs in the *Expression* "had to seem reasonable according to experience" (221) and therefore could be, implausibly, staged by an actor. So instead of building the case for their uncontaminated conditions of production as we now might expect, Darwin instead focused his discussion of photography around the context of reception including his readers; a canonic rhetorical concern.

9. Ekman, "Commentaries," in *E*, p. 21.

10. From the perspective of brain science, the current leading critique of Ekman has been developed by Lisa Feldman Barrett and colleagues. See for instance Maria Gendron, Debi Roberson, Jacoba Marietta van der Vyver, and Lisa Feldman Barrett, "Perceptions of Emotion from Facial Expressions Are Not Culturally Universal: Evidence from a Remote Culture," *Emotion* no. 14 (2014): 251–262.

11. See Ralph Adolphs, Daniel Tranel, and Antonio R. Damasio, "The Human Amygdala in Social Judgment," *Nature* 393, no. 6684 (1998): 470–474. These results are summarized in Damasio, *The Feeling of What Happens: Body and Emotion in the Making of Consciousness* (New York: Houghton Mifflin Harcourt, 1999). For an extended critique of Damasio, see Daniel M. Gross, *The Secret History of Emotion: From Aristotle's* Rhetoric *to Modern Brain Science* (Chicago: University of Chicago Press, 2006), 21–50.

12. Damasio, *The Feeling of What Happens*, 66–67.

13. Along similar lines see Paolo Legrenzi and C. A. Umiltà, *Neuromania: On the Limits of Brain Science* (Oxford: Oxford University Press, 2011); Nikolas S. Rose and Joelle M. Abi-Rached, *Neuro: The New Brain Sciences and the Management of the Mind* (Princeton: Princeton University Press, 2013).

14. Candace B. Pert, *Molecules of Emotion: Why You Feel the Way You Feel* (New York: Simon and Schuster, 1997).

15. Jerome Kagan, *What Is Emotion? History, Measures, and Meanings* (New Haven: Yale University Press, 2007), 23. Kagan's quarrel with the brain science of emotion in the tradition of Damasio/Ekman highlights the irreducible relevance of personal expectation. Kagan cites, for example, adults judging whether the facial expression of a male or female face was happy or fearful took longer to decide whether a man's face was fearful because they did not expect to see photos of fearful men (63). Also, like other brain structures the amygdala serves many functions, so when a scientist labels the amygdala's response to the unexpected appearance of faces with fearful expressions they must account for the response uncertainty created by the cognitive effort needed to classify the face or to understand why it was presented (70). It is impossible to create in the laboratory human emotions as they occur in "ecologically natural settings" summarizes Kagan (81), and therefore we have good reason to doubt conclusions about the latter that depend on evidence produced in the former. For an excellent and generously footnoted review of the Ekman controversy with special attention to the critique of Ekman within the natural and social scientific literature, see Ruth Leys, "How Did Fear Become a Scientific Object and What Kind of Object Is It?" *Representations* 110, no. 1 (Spring 2010): 66–104. A good summary article on the Ekman critique and the emotional turn in historical research is William Reddy, Barbara Rosenwein, and Peter Stearns, "The History of Emotions," interview by Jan Plamper, *History and Theory* 49, no. 2 (May 2010): 237–265.

16. The apt description of this book as "valedictorian" comes from an article on the positive psychology movement that includes some of Kagan's Harvard colleagues. See Sue Halpern, "Are You Happy?," a review of *The How of Happiness: A Scientific Approach to Getting the Life You Want* by Sonja Lyubomirsky; *Happier: Learn the Secrets to Daily Joy and Lasting Fulfillment* by Tal Ben-Shahar; *Stumbling on Happiness* by Daniel Gilbert; *Against Happiness: In Praise of Melancholy* by Eric G. Wilson; and *What Is Emotion?: History, Measures, and Meanings* by Kagan, *New York Review of Books,* April 3, 2008, www.nybooks.com/articles/archives/2008/apr/03/are-you-happy/.

17. Paul Ekman, "Basic Emotions," in *Handbook of Cognition and Emotion,* ed. Tim Dalgleish and Mick J. Power (Chichester, UK: John Wiley and Sons, 1999), 46.

18. Ekman, "Afterword: Universality of Emotional Expression? A Personal History of the Dispute," in *E,* 392; emphasis added. In 1999 Ekman offered this revision: "I propose . . . the following list of emotions: amusement, anger, contempt, contentment, disgust, embarrassment, excitement, fear, guilt, pride in achievement, relief, sadness/distress, satisfaction, sensory pleasure, and shame. When it is remembered that each of these words denotes a family of related emotions, then this list of 15 emotions is quite expanded. Clearly, it omits some affective phenomena which others have considered to be emotions. Guilt is a likely candidate, and I have no reason to make a guess one way or another." Paul Ekman, "Basic Emotions," 55.

19. Although I focus on CAL, a similar project is underway in art history and criticism. See especially David Freedberg, "Empathy, Motion, and Emotion," in *Wie sich Gefühle Ausdruck verschaffen: Emotionen in Nahsicht,* ed. Klaus Herding and Antje Krause-Wahl (Taunusstein, DE: Dreisen, 2007), 17–51, and Freedberg and Vittorio Gallese, "Motion, Emotion, and Empathy in Esthetic Experience," *Trends in Cognitive Sciences* 11, no. 5 (2007): 197–203. In Freedberg's work, like in some of the CAL material examined here, Darwin's complex understanding of emotional situations is restricted by placing the *Expression* in the Ekman/Damasio trajectory along with two other familiar touchpoints—primatologist Frans B. M. de Waal's work on bonobo empathy and the recent discovery of mirror neurons. Freedberg can thus reach the unlikely conclusion

that in Aby Warburg's *Pathosformel* "the outward forms of movement in a work revealed the inner emotions of the figure concerned" (Freedberg, "Empathy, Motion, and Emotion," 25). In fact the opposite is true: Warburg's *Pathosformel* is about the "accessory in motion" *(bewegtes Beiwerk)*, like the wind-hair draperies of Botticelli's *Primavera*, as a dynamic historical formula. For a discussion of this latter view, see Georges Didi-Huberman, "The Imaginary Breeze: Remarks on the Air of the Quattrocento," *Journal of Visual Culture* 2, no. 3 (2003): 275–289.

20. Quoting Karen Sainchez-Eppler: June Howard, "What Is Sentimentality?" *American Literary History* 11, no. 1 (Spring 1999): 63–81, at 64. The trendiness of CAL is evident in Patricia Cohen, "The Next Big Thing in English: Knowing They Know That You Know," *New York Times*, March 31, 2010, C1, and readers' comments published online as "Can 'Neuro Lit Crit' Save the Humanities?" community.nytimes.com/comments/roomfordebate.blogs.nytimes.com/2010/04/05/can-neuro-lit-crit-save-the-humanities/.

21. See Mary Thomas Crane, *Shakespeare's Brain: Reading with Cognitive Theory* (Princeton: Princeton University Press, 2001), 4; hereafter abbreviated *SB*.

22. See Jane F. Thrailkill, *Affecting Fictions: Mind, Body, and Emotion in American Literary Realism* (Cambridge, MA: Harvard University Press, 2007), 250.

23. Suzanne Keen, *Empathy and the Novel* (Oxford: Oxford University Press, 2007), 27. See also G. Gabrielle Starr, *The Neuroscience of Aesthetic Experience* (Cambridge, MA: MIT Press, 2013).

24. When Terence Cave writes about "literary affordances"—how to sit on a chaise longue in Jonathan Franzen's novel *The Corrections* (46)—he is onto something similar. In his compelling example, literature foregrounds in a particular way, and formalizes, an experience such that its status as fiction makes no difference phenomenologically (129). Like me, Cave is interested in how literature helps compose our environments (3). My reservation about Cave's approach can be attributed to the weight he places on cognition per se, as when he writes that "humans live in a cognitive ecology" (5). The statement is not untrue. But it unduly narrows the analytic frame like any other such not-untrue statement, such as "humans live in an emotional ecology," or even "the planet is 71% water." In any case the ecology analyzed will then be predictably cognitive, or emotional, or watery, or what have you. That is to say after such a statement, one subsequently makes assumptions that are precisely at issue in terms of both the phenomenology, and the rhetoric. See Terence Cave, *Thinking with Literature: Toward a Cognitive Criticism* (Oxford: Oxford University Press, 2016).

25. Jonathan Kramnick, "An Aesthetics and Ecology of Presence," *European Romantic Review* 26, no. 5 (2015): 315–327. Kramnick references Gibson and Noë in his reading of eighteenth-century local-descriptive poetry, and the fiction of Sterne. Kramnick's pursuit of the aesthetics of presence "with its emphasis on skilled action and its embrace of naïveté" (315) contrasts in some ways the approach I pursue in this project. I consider some critical perspective figured in Sterne, Equiano, and Austen an essential part of the perceptual environments at issue in each case. Kramnick's book-length project is tentatively titled *Paper Minds: Literature and Some Problems of Consciousness.*

26. Though distinct from extended-mind philosophers who typically cite artificial intelligence research, not linguistic research, practitioners of "cognitive rhetoric" drawing on the work of Lakoff, Turner, et al. have produced some interesting lexical analyses of literature, and a perspective in some ways sympathetic with mine insofar as it would endorse this passage from Turner: "What can be recruited to mental work depends on social and cultural location. Parts of the repertoire are common and can be assumed for any audience, while other parts are special

to special communities or special situations. Consequently, it is a basic principle of rhetorical theory that what works in one situation may not work in another. One of Aristotle's definitions of rhetoric is: "the mental ability to see the available means of persuasion *in any particular situation*." Mark Turner, "Toward the Founding of Cognitive Social Science," *Chronicle of Higher Education*, October 5, 2001, markturner.org/checss.html. The foundational book is Turner, *Reading Minds: The Study of English in the Age of Cognitive Science* (Princeton: Princeton University Press, 1991).

27. Thrailkill, *Affecting Fictions*, 51 and 53.

28. Ibid., 15.

29. Lauren Berlant, "Introduction: Compassion (and Witholding)," in *Compassion: The Culture and Politics of an Emotion*, ed. Berlant (London: Routledge, 2004), 3.

30. The patron saint of universal emotion, Adam Smith, gives us this charming formulation: a sudden fall into disease may deserve sympathy, whereas being jilted by one's mistress or henpecked by one's wife doesn't. See Adam Smith, *The Theory of Moral Sentiments* (Amherst, NY: Prometheus, 2000 [1759]), 59.

31. Virgil, *The Aeneid*, trans. Frederick Ahl (Oxford: Oxford University Press, 2007), 2:223; 35.

32. In 1767 Henri Fusseli translates Winckelmann on expression: "The last and most eminent characteristic of the Greek works is a noble simplicity and sedate grandeur in Gesture and Expression. . . .'Tis in the face of Laocoon this soul shines with full lustre, not confined however to the face, amidst the most violent sufferings. Pangs piercing every muscle, every labouring nerve; pangs which we almost feel ourselves, while we consider—not the face, nor the most expressive parts—only the belly contracted by excruciating pains" (J. J. Winckelmann, *Reflections on the Painting and Sculpture of the Greeks*, trans. Henri Fusseli [London, 1765], 30).

33. Darwin's discussion of artistic expression footnotes Lessing's *Laocoön* (see *E*, 436n16). I thank William Schupbach of the Wellcome Library for the History and Understanding of Medicine for notes on Laocoön and the history of pain.

34. See Charles Bell, *The Anatomy and Philosophy of Expression as Connected with the Fine Arts* (London, 2006 [1844]), 174–175.

35. See note 19 above on Aby Warburg's *Pathosformel* and the skillfully given accessory.

36. It is worth noting that the threatening snake has been considered the quintessential "emotionally competent object" in the tradition running from Descartes's *The Passions of the Soul* through the brain science of LeDoux and Damasio.

37. David Chalmers, Foreword to Andy Clark, *Supersizing the Mind: Embodiment, Action, and Cognitive Extension* (Oxford: Oxford University Press, 2008), xiv. As well as works cited in the body of this chapter, see Michael Wheeler, *Reconstructing the Cognitive World: The Next Step* (Cambridge, MA: MIT Press, 2005); Evan Thompson, *Mind in Life: Biology, Phenomenology, and the Sciences of Mind* (Cambridge, MA: Harvard University Press, 2007); Douglas Robinson, *Feeling Extended: Sociology as Extended Body-Becoming-Mind* (Cambridge, MA: MIT Press, 2013); Giovanna Colombetti, *The Feeling Body: Affective Science Meets the Enactive Mind* (Cambridge, MA: MIT Press, 2014). To date, the most developed critique of the situated cognition movement and its varieties—embodied, embedded, extended—can be found in Frederick Adams and Kenneth Aizawa, *The Bounds of Cognition* (Malden, MA, 2008). For a negative review see Sven Walter and Miriam Kyselo, review of *The Bounds of Cognition* by Adams and Aizawa, *Erkenntnis* 71 (September 2009): 277–281.

38. Chalmers, Foreword, xiv.

39. Hubert L. Dreyfus, "Why Heideggerian AI Failed and How Fixing It Would Require Making It More Heideggerian," *Artificial Intelligence* 171, no. 18 (2007): 1137–1160, at 1160, emphasis original.

40. For some related criticism see also Aileen Douglas, *Uneasy Sensations: Smollett and the Body* (Chicago: University of Chicago Press, 1995), especially Chapter 2, "'Matters out of Place': *Travels through France and Italy*." Douglas writes about how "Dirt, in the *Travels*, is the manifestation through which the narrator confronts his own sense of dislocation and bewilderment" (27); also "discomfort and unease" are major themes in the *Travels* (28).

41. A good summary can be found in Steven Pinker, "Toward a Consilient Study of Literature," *Philosophy and Literature* 31, no. 1 (2007): 162–178. In fact Pinker first uses cheesecake to talk about music, not the novel, in *How the Mind Works* (New York: W. W. Norton, 1997), 524–525. It is Joseph Carroll's review of Pinker's *How the Mind Works* (*Philosophy and Literature* 22, no. 2 [1998]: 478–485) that draws this particular connection.

42. Relevant here is Alenka Zupančič on situation comedy. Via Hegel and Lacan, she provides a theory of comedy defined by situations run amok (impossible locations) vs. tragedy, which focuses on the experience of dislocation. Examples: the joke about the chicken thinking you are a seed; Charlie Chaplin, *Modern Times*. This taxonomy would suggest, accurately I think, that Sterne exploits the tragicomic possibilities of dislocation. Alenka Zupančič, *The Odd One In: On Comedy* (Cambridge, MA: MIT Press, 2008).

43. Lisa Zunshine, *Why We Read Fiction: Theory of Mind and the Novel* (Columbus: Ohio State University Press, 2006), 47.

44. Terence Cave considers a privative like "leafless" to be a miniature instance of the rhetorical figure *praeteritio*: "mentioning something while simultaneously saying you're not going to mention it" (*Thinking with Literature*, 90).

45. Laurence R. Horn, *A Natural History of Negation* (Chicago: University of Chicago Press, 1989). Cited edition is republished (Stanford: CSLI, 2001). Analytically shared between Horn and Cave is affinity with the work of cognitive pragmatist and language philosopher Paul Grice.

46. Jane Austen, James Kinsley, and Adela Pinch, *Emma* (Oxford: Oxford University Press, 2003), 321.

47. For a summary of the debate see Hubert L. Dreyfus, "Heidegger's Critique of the Husserl/Searle Account of Intentionality," *Social Research* 60, no. 1 (1993): 17–38.

48. Kenneth Burke, the rhetorician and literary critic, observed something similar about eighteenth-century bourgeois poetry of the sublime as it appeared from a perspective in the following century; see the discussion in my introduction. Burke is making a point about how we must read the rhetoric of emotion historically. We must always ask who is attributing which emotions to whom—bracket sincerity—and what this act of attribution is supposed to accomplish. In eighteenth-century sentimental literature, including some works of what Burke calls the poetry of the sublime, authorial attribution of emotion, e.g., the "man of feeling" (like Yorick) says little about Yorick's sincere emotions whatever that might mean, and quite a lot about the "discomfort" that would be addressed by this very attribution. One added virtue of Sterne's work is that he provides at the same time a kind of reflective meta-commentary that makes just this point.

49. Joan Roughgarden, *Evolution's Rainbow: Diversity, Gender, and Sexuality in Nature and People* (Berkeley: University of California Press, 2009 [2004]), 5.

50. See ibid., 177.

51. See ibid., 311.

52. Prodger discusses at length the first of these images, best known as *Ginx's Baby*. Though

considered a technical tour de force because Rejlander had supposedly captured body and facial muscles in vigorous movement, the picture was not a true photograph but a drawing made to look like an original photograph. Among other things, Prodger discusses how, by adding details like a large padded chair to support the crying baby, Rejlander's "corrections" convey "a tangible moment in everyday life readers could mentally verify, and with which they could sympathize" (Prodger, *Darwin's Camera*, 125).

Chapter Two

1. Hereafter I will cite this Penguin Classics edition internally. Olaudah Equiano, *The Interesting Narrative and Other Writings*, ed. with an introduction and notes by Vincent Carretta (New York: Penguin, 2003), 77.

2. Maurice Merleau-Ponty, *Phenomenology of Perception*, trans. Colin Smith (London: Routledge, 1989 [1945]), 136.

3. Since 1995, Penguin Classics editor and Equiano biographer Vincent Carretta has made a controversial claim that Equiano was in fact born in South Carolina, which would mean that his accounts of life in Africa and of the Middle Passage are not eyewitness. A 2006 article by Cathy N. Davidson summarizes the controversy up to that point as it focuses on two key documents—a 1759 parish baptismal record and a 1773 ship muster—while providing a rebuttal that focuses on the unreliability of these two documents. She makes the case for Equiano's basic credibility within the context of the *Narrative*'s original publication. Cathy N. Davidson, "Olaudah Equiano, Written by Himself," *Novel* 40, no. 1–2 (2006): 18–51. That I find Davidson's argument persuasive is immaterial to my work in this chapter, which is about the rhetoric of emotion.

4. Orlando Patterson, *Slavery and Social Death: A Comparative Study* (Cambridge, MA: Harvard University Press, 1985).

5. Carretta, Introduction, xix; Henry Louis Gates Jr., *Black Imagination and the Middle Passage* (Oxford: Oxford University Press, 1999), 55.

6. Emmanuel Kant, *Critique of Pure Reason*, trans. Norman Kemp Smith (New York: Saint Martin's, 1929), 138.

7. The classic work on Du Bois and European idealism from Kant to Hegel is Paul Gilroy, *The Black Atlantic: Modernity and Double Consciousness* (Cambridge, MA: Harvard University Press, 1993).

8. Frantz Fanon, *Black Skin, White Masks* (New York: Grove Press, 2008 [1952]).

9. And where, according to Young, the lived body is a "unified idea of a physical body acting and experiencing in a specific socio-cultural context; it is body-in-situation" (16). Feminine bodily existence "stands in discontinuous unity with both itself and its surroundings" (38). Iris Marion Young, *On Female Body Experience: "Throwing like a Girl" and Other Essays* (Oxford: Oxford University Press, 2005). A critique of the ways in which Merleau-Ponty's model of sexuality presumes a general or universal orientation toward the world is found in Judith Butler, "Sexual Ideology and Phenomenological Description: A Feminist Critique of Merleau-Ponty's *Phenomenology of Perception*," in *The Thinking Muse: Feminism and Modern French Philosophy*, ed. Jeffner Allen and Iris Marion Young (Bloomington: Indiana University Press, 1989), 85–100.

10. Salamon writes about a young transman named Aidan: "his internal sense of discomfort with his external appearance may not be a simple misalignment or mismatch, but an internal sense of dysphoria that becomes amplified as it circuits from his body to the gaze of an external world that is brutally hostile to gender ambiguity to become internalized and incorporated as a

part of his gendered sense of self." Gayle Salamon, *Assuming a Body: Transgender and Rhetorics of Materiality* (New York: Columbia University Press, 2010), 117. See also my following chapter 3 on hostile environments.

11. Fred Moten, "Knowledge of Freedom," *CR: New Centennial Review* 4, no. 2 (2004): 269–310, at 299. Moten's figure of the "consuming ship" appears nicely antithetical to a figure of the ship-as-womb, a comfortable environment sustaining life amidst the waters.

12. Hence my approach is close to Sara Ahmed's "queer phenomenology." Taking seriously the "orientation" in "sexual orientation," Ahmed develops in her book of the same name a beautiful rereading of the phenomenological tradition after Husserl. In the introduction she notes, for instance, how phenomenology "is full of queer moments; as moments of disorientation that Maurice Merleau-Ponty suggests involve not only 'the intellectual experience of disorder, but the vital experience of giddiness and nausea, which is the awareness of our contingency, and the horror with which it fills us'" (4; 138–139 on a project that begins with disorientation). Ahmed, instead of accounting for how these moments are overcome as bodies become reoriented, wants to stay with such moments so we might "achieve a different orientation toward them . . . find joy and excitement in the horror." Of course I'm not trying to find joy and excitement in Equiano's horror, but rather I'm trying to demonstrate how the literary humanities must be brought to bear in a case like this. See Sara Ahmed, *Queer Phenomenology: Orientations, Objects, Others* (Durham: Duke University Press, 2006).

13. Cathy N. Davidson calls the most consistent trope structuring the plot of *The Interesting Narrative* "the existential rug-pull" (20), and she explains how this trope produces a hybrid form of narrative. "Given the inherent insecurity of being an African or of African descent in a world where black people are enslaved, watchfulness is a precondition of Equiano's existence. His is not a Moses-like journey from slavery to freedom, but rather, an episodic and often anxious narrative. . . . The hybrid form of *The Interesting Narrative* replicates the profound uncertainty of the narrator. The text combines (in unequal parts) slave narrative, sea yarn, military adventure, ethnographic reportage, historical fiction, travelogue, picaresque saga, sentimental novel, allegory, tall tale, pastoral origins myth, gothic romance, conversion tale, and abolitionist tract, with different features coming to the fore at different times, and the mood vacillating accordingly." Cathy N. Davidson, "Olaudah Equiano, Written by Himself," 19.

14. In *Our Aesthetic Categories: Zany, Cute, Interesting*, Sianne Ngai spends time with Sterne before she notes of the interesting: "its desire to know reality by comparing one thing with another, or by lining up what one realizes one doesn't know against what one knows already" (6). In contrast to this epistemic logic, my analysis of Equiano's *Interesting Narrative* draws out a very different narrative logic not destined for realism, but rather a kind of antirealism that makes room for sacred moments in a patient narrative that must in fact reach some acceptable level of plausibility. Political theology also distinguishes my approach to this eighteenth-century narrative material from recent work on queer temporality, which does have some similar analytic concerns. Notably Lee Edelman, *No Future: Queer Theory and the Death Drive* (Durham: Duke University Press, 2004); Judith Halberstam, *In a Queer Time and Place: Transgender Bodies, Subcultural Lives* (New York: New York University Press, 2005); Elizabeth Freeman, *Time Binds: Queer Temporalities, Queer Histories* (Durham: Duke University Press, 2010).

15. Daniel Kahneman and Dale T. Miller, "Norm Theory: Comparing Reality to Its Alternatives," *Psychological Review* 93, no. 2 (1986): 136–153, at 145. Emotional amplification is configured by Kahneman within the new science of hedonic psychology.

16. P. Brickman, R. Janoff-Bulman, and D. Coates, "Lottery Winners and Accident Victims: Is Happiness Relative?" *Journal of Personality and Social Psychology* 36, no. 8 (1978): 917–927. The

hedonic treadmill was introduced by Brickman and Campbell, "Hedonic Relativism in Planning the Good Society," in *Adaption-Level Theory: A Symposium*, ed. M. H. Apley (New York: Academic, 1971), 287–302. Daniel Kahneman, Ed Diener, and Norbert Schwarz, eds., *Well-Being: The Foundations of Hedonic Psychology* (New York: Sage, 1999) is the foundational 1999 textbook. In the first chapter of this book (13), Kahneman qualifies and frames this material on the hedonic treadmill. Subsequent in-text citations are from this book.

17. A summary of the literature on psychological adaptation and the example of incarceration can be found in Shane Frederick and George Loewenstein, "Hedonic Adaptation," in *Well-Being: Foundations of Hedonic Psychology*, ed. Daniel Kahneman, Ed Diener, and Norbert Schwarz (New York: Russell Sage Foundation, 1999), 311–312.

18. Daniel Kahneman, Peter P. Wakker, and Rakesh Sarin, "Back to Bentham? Explorations of Experienced Utility," *Quarterly Journal of Economics* 112, no. 2 (1996): 375–376.

19. From the title of Dan Ariely's *Predictably Irrational: The Hidden Forces That Shape Our Decisions* (New York: Harper Perennial, 2008). Another way to put this is hedonic psychologists are interested in the psychology of systematic errors, for example the systematic misprediction of one's own future feelings ("I can eat just one chip," "I will always love you"). Gregory Loewenstein and David Schkade, "Wouldn't It Be Nice? Predicting Future Feelings," in *Well-Being*, ed. Daniel Kahneman, Ed Diener, and Norbert Schwarz, 85–86.

20. For an extensive historical treatment of this topic see Frank Biess and Daniel M. Gross, eds., *Science and Emotion after 1945: A Transatlantic Perspective* (Chicago: University of Chicago Press, 2014); also Paul Erickson, Judy L. Klein, Lorraine Daston, Rebecca Lemov, Thomas Sturm, and Michael D. Gordin, *How Reason Almost Lost Its Mind: The Strange Career of Cold War Rationality* (Chicago: University of Chicago Press, 2013); Barbara Maria Stafford, ed., *Bridging the Humanities-Neuroscience Divide: A Field Guide to a New Meta-Field* (Chicago: University of Chicago Press, 2011).

21. Daniel Kahneman, *Thinking, Fast and Slow* (New York: Farrar, Straus & Giroux, 2011) is the post-Nobel megahit that made Kahneman a household name amongst intellectuals, and provides an updated account of cognitive science notably in the research trajectory of Joseph LeDoux and colleagues. For a critique of LeDoux's brain science of emotion see Daniel M. Gross, *The Secret History of Emotion*, chapter 1.

22. The OED gives us "a little reminder of home, which may happily make you homesick."

23. Arthur A. Stone, Saul S. Schiffman, and Martin W. DeVries, "Ecological Momentary Assessment," in *Well-Being*, ed. Daniel Kahneman, Ed Diener, and Norbert Schwarz, 28.

24. Alva Noë, *Varieties of Presence* (Cambridge, MA: Harvard University Press, 2012).

25. Alva Noë in one of his interviews around the book *Out of Our Heads*: Retrieved March 5, 2014. http://edge.org/conversation/life-is-the-way-the-animal-is-in-the-world. Similarly, Alva Noë on experience: "Experience is not something that happens in us. It is something we do. Experience itself [is] a kind of dance—a dynamic of involvement and engagement with the world around us." *Varieties of Presence* (Cambridge, MA: Harvard University Press, 2012), 130.

26. Vincent Carretta, *Equiano, the African: Biography of a Self-Made Man* (Athens: University of Georgia Press, 2005), 98.

27. This is a central problematic for Judith Butler. See especially her work on "passionate attachments" in the introduction to *The Psychic Life of Power: Theories of Subjection* (Stanford: Stanford University Press, 1997). In this work Butler is interested in a scenario where the desire for subjection is not simply "love of the shackles," but rather the desire to persist in one's own being, as it is passionately attached to the power that subjects—hence ambivalence (27–28). Equiano's passionate "attachment" to his master exemplifies this dynamic, I think, as it is

structured by economic and not just political relations: Equiano's attachments are a function of investment or lack thereof.

28. For example, Hubert Dreyfus, "Intelligence without Representation—Merleau-Ponty's Critique of Mental Representation," *Phenomenology and the Cognitive Sciences* 1, no. 4 (2002): 367–383, at 378. "According to Merleau-Ponty, in absorbed, skillful coping, I don't need a mental representation of my goal. Rather, acting is experienced as a steady flow of skillful activity in response to one's sense of the situation. Part of that experience is a sense that when one's situation deviates from some optimal body-environment relationship, one's activity takes one closer to that optimum and thereby relieves the 'tension' of the deviation. One does not need to know, nor can one normally express, what that optimum is. One's body is simply solicited by the situation to get into equilibrium with it. As Merleau-Ponty puts it: 'Whether a system of motor or perceptual powers, our body is not an object for an "I think," it is a grouping of lived-through meanings which moves towards its equilibrium' (Merleau-Ponty 1962, 153)." An update of this discussion around pragmatic Merleau-Ponty can be found in Donnchadh O'Conaill, "On Being Motivated," *Phenomenology and the Cognitive Sciences* 12, no. 4 (2013): 579–595.

29. In *Queer Phenomenology* Ahmed draws attention to the word "comfort." "The word 'comfort' suggests well-being and satisfaction, but it also suggests an ease and an easiness. Comfort is about an encounter between more than one body, which is the promise of a 'sinking' feeling. To be comfortable is to be so at ease with one's environment that it is hard to distinguish where one's body ends and the world begins. One fits, and in the active fitting, the surface of bodies disappear from view. White bodies are comfortable *as they inhabit spaces that extend their shape.* The bodies in spaces 'point' toward each other, as a 'point' that is not seen as it is also 'the point' from which we see. In other words, whiteness may function as a form of public comfort *by allowing bodies to extend into spaces that have already taken their shape.* Those spaces are lived as comfortable as they allow bodies to fit in; the services of social space are already impressed upon by the shape of such bodies" (134). So "you can feel the categories that you fail to inhabit: they are sources of discomfort," explains Ahmed, while at the same time this failure is dynamic; "discomfort . . . allows things to move" (154).

30. I use the Findlay translation with hypertext to his commentary and to the original German, as it appears in Hegel by Hypertext. http://www.marxists.org/reference/archive/hegel/index.htm.

31. At this point Hegel has reemerged as relevant to the philosophy and science of mind, in particular the "extended mind" hypothesis. See instances in Anthony Crisafi and Shaun Gallagher, "Hegel and the Extended Mind," in *AI & Society* 25, no. 1 (2010): 123–129; and Douglas Robinson, *Feeling Extended: Sociality as Extended Body-Becoming-Mind* (Cambridge, MA: MIT Press, 2013). Hegel has the idea that "the mind is expressed in social institutions," and he has "a clearly elaborated understanding of the situated embodiment of mind, of mind arising out of the body's interactions with its environment" (35). They all cite Hegel's "philosophy of mind" section of his *Encyclopedia of the Philosophical Sciences*: "Human expression includes, e.g., the upright figure in general, the formation especially of the hand, as the absolute tool, of the mouth, laughter, weeping, etc., and the spiritual tone diffused over the whole, which at once announces the physical body as the externality of a higher nature" (quoted in Robinson, 36–37).

32. Referencing Hegel's *The System of Ethical Life*, Robinson shows how the extended mind of Andy Clark contrasts with Hegel's extended mind alienation: "extending mind to tools for Hegel doesn't just help us think better—Clark's (2008) recurring refrain. It does that too, but it also blinds our thinking, makes it (and us) more tool-like" (42).

33. Hence the Marx/Engels critique of Feuerbach: "Even the objects of the simplest 'sen-

suous certainty' [like the cherry tree] are only given him through social demands, industry and commercial intercourse." Cited in Ahmed, *Queer Phenomenology*, 41. Like Lauren Berlant, Ahmed also draws attention to the way in which objects are aspirational; we can have an "objective" (56). Those aspirations can feel, moreover, compulsory as in finding an appropriate love object; we thus tend toward some objects and not others (85) while attributing volition to oneself inaccurately.

34. Objects appear queer (retinal image inversion experiment). "Finding out how a normal situation is restored amounts than to finding out how the new image of the world and one's own body can cause the other to 'pale', or 'displace' it. It is noticeable that the normal situation is the more successfully achieved in proportion as the subject is more active" (245, 289).

35. Merleau-Ponty is not *just* a Hegelian in the Marxist line of interpretation. Significant for our purposes, for instance, is phenomenology that goes back to Heidegger, and then gets picked up by our philosophers interested in "the background of intentionality," i.e., the tacit or what must be taken for granted, in order to make room for an intentional act like when I take a room full of furniture as a whole, and deal with it by sitting down in a chair. Dreyfus points to Heidegger and his work on practical, everyday orientation that makes action possible and being there, Dasein, as such. Merleau-Ponty, in turn, writes about "the zone of not being in front of which precise beings, figures and points can come to light" (100), suggesting the ways in which *not being* has its own relational phenomenology beyond the Hegelian account of being negated. The approach of Heidegger and Merleau-Ponty allows for ontological analysis of a hammer, for instance, not available in such detail to Hegel, who can treat the hammer only at more abstract levels of thinghood.

36. Michael Argyle, "Causes and Correlates of Happiness," in *Well-Being*, ed. Daniel Kahneman, Ed Diener, and Norbert Schwarz, 370. "The main explanation for the lower happiness and self-esteem of members of ethnic minorities is that they have lower incomes, less education, and less skilled jobs. The effect of ethnicity when these have been controlled is quite small" (362).

37. "The design or intention of the author is neither available nor desirable as a standard for judging the success of a work of literary art." William K. Wimsatt and Monroe C. Beardsley, "The Intentional Fallacy," *Sewanee Review* 54, no. 3 (1946): 468–488. Revised and republished in *The Verbal Icon: Studies in the Meaning of Poetry* (Lexington: University of Kentucky Press, 1954), 3–18.

38. Thanks to a note from Carretta, 240n36.

39. Saidiya V. Hartman, *Scenes of Subjection: Terror, Slavery, and Self-Making in Nineteenth-Century America* (Oxford: Oxford University Press, 1997).

40. "Positive emotions signify that life is going well, the person's goals are being met, and resources are adequate (e.g., Cantor et al., 1991; Carver & Scheier, 1998; Clore, Wyer, Dienes, Gasper, & Isbell, 2001). In these circumstances, as Fredrickson (1998, 2001) has so lucidly described, people are ideally situated to 'broaden and build.' In other words, because all is going well, individuals can expand their resources and friendships; they can take the opportunity to build their repertoire of skills for future use; or they can rest and relax to rebuild their energy after expending high levels of effort. Fredrickson's model (Fredrickson, 2001) suggests that a critical adaptive purpose of positive emotions is to help prepare the organism for future challenges." Sonja Lyubomirsky, Laura King, and Ed Diener, "The Benefits of Frequent Positive Affect: Does Happiness Lead to Success?" *Psychological Bulletin* 131, no. 6 (2005): 803–855, at 804.

41. Elsewhere Noë on bearing: "This is the key to our puzzle. What governs the character of our experience—what makes experience the kind of experience it is—is not the neural activity in our brains on its own; it is, rather, our ongoing dynamic relation to objects, a relation that, as

in this case, clearly depends on our neural responsiveness to changes in our relation to things. It is there, in this extended, sensorimotor involvement with the world, that we find the resources to explain why we see when we use the tactile-visual sensory substitution system. My proposal, specifically, is this: We *see* with Bach-y-Rita's system because the relationship that system sets up and maintains between the perceiver and the object is, in ways that can be made precise, the sort of relation that we bear to things when we see them. What causes the effects for consciousness of neural activity in the touch-dedicated parts of the brain to change? Answer: the world in our relation to it" (*Out of Our Heads*, 59).

42. A good introduction can be found in Jared Sexton, "The Social Life of Social Death: On Afro-Pessimism and Black Optimism," *InTensions Journal* 5 (2011): 1–47, at 28: "Nothing in afro-pessimism suggests that there is no black (social) life, only that black life is not social life in the universe formed by the codes of state and civil society, of citizen and subject, of nation and culture, of people and place, of history and heritage, of all the things that colonial society has in common with the colonized, of all that capital has in common with labor—the modern world system. Black life is not lived in the world that the world lives in, but it is lived underground, in outer space." Key figures also include Frank Wilderson and Fred Moten.

43. Vincent Carretta's note: Adapted from "The Spiritual Victory," No. 312 in Augustus Montague Toplady (1740–1778), *Psalms and Hymns for Public and Private Worship. Collected (for the Most Part), and Published, by August Toplady* (London, 1776).

44. Mary Wollstonecraft, *The Analytical Review* Vol. 4 (London: Printed for J. Johnson, 1789), 27–29.

Chapter Three

1. D. T. Max, "The Literary Darwinists," *New York Times*, November 6, 2005. Also notably Denis Dutton and publicity around *The Art Instinct*; positive reviews of Brian Boyd's *On the Origin of Stories* in *New Scientist*, *Boston Globe*, etc. Terry Eagleton writes in *London Review of Books* September 24, 2009: "Brian Boyd's *On the Origin of Stories* . . . might well be a straw in the wind blowing contemporary criticism back from Culture to Nature." When it comes to the popularity of Cognitive Approaches to Literature, see for instance Patricia Cohen, "Next Big Thing in English: Knowing They Know That You Know," *New York Times*, March 31, 2010.

2. Nussbaum's controversial "capability approach" to development, worked out initially with Amartya Sen, unfolded next to her work on emotion, which is essentially neo-Stoic, with updates and scientific scaffolding by way of appraisal theories and their foundations in the brain science of Antonio Damasio and others. Elements of a cognitive approach to literature appear in Nussbaum's *Upheavals of Thought: The Intelligence of Emotions* (Cambridge: Cambridge University Press, 2001). For social policy implications see *Creating Capabilities: The Human Development Approach* (2011) and *Political Emotions: Why Love Matters for Justice* (2013), which mobilizes the brain science of Joseph LeDoux and especially the primatology of Frans de Waal on empathy and compassion. For Sen's rejection of a canonical list of essential capabilities à la Nussbaum, see "Human Rights and Capabilities," in the *Journal of Human Development* 6, no. 2 (2005): 151–166: "My own reluctance to join the search for such a canonical list arises partly from my difficulty in seeing how the exact lists and weights would be chosen without appropriate specification of the context of their use (which could vary), but also from a disinclination to accept any substantive diminution of the domain of public reasoning" (157).

3. Blakey Vermeule, "Critical Response VI: Wit and Poetry and Pope, or The Handicap Principle," *Critical Inquiry* [hereafter *CI*] 38, no. 2 (2012): 427.

4. Lisa Zunshine, ed., *Introduction to Cognitive Cultural Studies* (Baltimore: Johns Hopkins University Press, 2010).

5. Joseph Carroll, "Human Life History and Gene-Culture CoEvolution: An Emerging Paradigm," *Evolutionary Review: Art, Science, Culture* 2 (2011): 23–37, at 37.

6. Lisa Zunshine, ed., *The Oxford Handbook of Cognitive Literary Studies* (Oxford: Oxford University Press, 2015), 2.

7. Jerry Fodor, "Against Darwinism," *Mind & Language* 23, no. 1 (2008): 1–24, at 23.

8. *The Feeling of What Happens*, 125.

9. In this way Zunshine's volume can begin with New Left literary critic Raymond Williams, who uses the term centrally in *The Long Revolution*; quoted in Zunshine, 6–7.

10. Steven Pinker, "Toward a Consilient Study of Literature," *Philosophy and Literature* 31, no. 1 (2007): 162–178, here 176.

11. Lisa Zunshine, *Introduction to Cultural Cognitive Studies*, 62, quoting Spolsky, "Cognitive Literary Historicism," 168.

12. James J. Gibson, *The Ecological Approach to Visual Perception* (New York: Psychology Press, 1986 [1979]), 9.

13. In practice, Zunshine's ToM seems less than a cognitive universal and more like the particular way of reading discussed over the past decades by critics interested in how novel reading calibrates a certain kind of bourgeois subjectivity where, amongst other things, psychological interiority is formally privatized, and where mind reading becomes newly important as an exercise in market predictability. See for instance Deidre Shauna Lynch, *The Economy of Character: Novels, Market Culture, and the Business of Inner Meaning* (Chicago: University of Chicago Press, 1998).

14. Elaine Scarry, *Dreaming by the Book* (New York: Farrar, Straus & Giroux, 1999).

15. A controversial claim. For a different perspective, see for instance Garrett Stewart, *Reading Voices: Literature and the Phonotext* (Berkeley: University of California Press, 1990); and more recently Sarah Gleeson-White, "Auditory Exposures: Faulkner, Eisenstein, and Film Sound," *PMLA* 128, no. 1 (2013): 87–100.

16. George Lakoff, *Women, Fire, and Dangerous Things* (Chicago: University of Chicago Press, 1987), 215–216. Here Lakoff refers to a passage like this from Gibson: "To sum up, the characteristics of an environmental medium are that it affords respiration or breathing; it permits locomotion; it can be filled with illumination so as to permit vision; it allows detection of vibrations and detection of diffusing emanations; it is homogenous; and finally, it has the absolute axis of reference, up-and-down. All these offerings of nature, these possibilities or opportunities, these *affordances* as I will call them, are invariant. They have been strikingly constant throughout the whole evolution of animal life." Gibson, *The Ecological Approach*, 14–15.

17. Andrea Scarantino, "Affordances Explained," *Philosophy of Science* 70, no. 5 (2003): 949–961, at 950.

18. Bitzer's definition: "Any exigence is an imperfection marked by urgency; it is a defect, an obstacle, something waiting to be done, a thing which is other than it should be. In almost any sort of context, there will be numerous exigences, but not all are elements of a rhetorical situation—not all are rhetorical exigences." Lloyd F. Bitzer, "The Rhetorical Situation," *Philosophy and Rhetoric* 1, no. 1 (1968): 1–14, at 6.

19. Before Bitzer on constraints, Kenneth Burke in the 1930s wrote about "recalcitrances" that set propositional thought against assumptive backgrounds including structures of prohibition, impossibility, and resistance in the real world. So for example "I am a bird" must appear as a poetic expression because it defies at the very least physical laws. Kenneth Burke, *Permanence and Change: An Anatomy of Purpose*, 3rd ed. (Berkeley: University of California Press, 1984),

255. For a good summary article see Lawrence J. Prelli, Floyd D. Anderson, and Matthew T. Althouse, "Kenneth Burke on Recalcitrance," *Rhetoric Society Quarterly* 41, no. 2 (2011): 97–124.

20. S. Kitayama, B. Mesquita, and M. Karasawa, "Cultural Affordances and Emotional Experience: Socially Engaging and Disengaging Emotions in Japan and the United States," *Journal of Personality and Social Psychology* 91, no. 5 (2006): 890–903.

21. Lawrence Kaufman and Fabrice Clément, "How Culture Comes to Mind: From Social Affordances to Cultural Analogies," *Intellectica* 2–3, no. 46/47 (2007): 221–250.

22. See also Nico H. Frijda, *The Emotions* (Cambridge: Cambridge University Press, 1986), 189–190: "Roughly speaking, emotional experience corresponds to a motive appearance involving 'demand characteristics' (Lewin 1937). Demand characteristics, again roughly speaking, refer to the 'valence' (again Lewin's) of the situation as a whole: whether it is attractive or aversive or merely 'demanding.' . . . We will refer to such a motive appearance of situations as *situational meaning structures.*"

23. Erik Rietveld, "Situated Normativity: The Normative Aspect of Embodied Cognition in Unreflective Action," *Mind* 117, no. 468 (2008): 973–1001, at 987.

24. Hence also the field Social Psychology, which is deeply indebted to Lewin. See for example the classic introductory work by Lee Ross and Richard E. Nisbett, *The Person and the Situation: Perspectives of Social Psychology* (New York: McGraw-Hill, 1991) with a new introduction by Malcolm Gladwell (London: Pinter and Martin, 2011).

25. "Muslim woman wins $5 million verdict from AT&T for discrimination," http://www.reuters.com/article/us-usa-court-muslim-att-idUSBRE8431P520120504.

26. Ellen Spolsky, "Making 'Quite Anew': Brain Modularity and Creativity," in Zunshine, *Introduction to Cognitive Cultural Studies*, 84.

27. Kim Sterelny, *Thought in a Hostile World: The Evolution of Human Cognition* (Oxford: Blackwell, 2003).

28. Patrick Colm Hogan on "Literary Universals" in Zunshine, *Introduction to Cognitive Cultural Studies* (50), cites the universal "We don't like to laugh at hostile jokes directed at groups of which we are part."

29. Gordon H. Orians and Judith H. Heerwagen, "Evolved Responses to Landscapes," in *The Adapted Mind: Evolutionary Psychology and the Generation of Culture*, ed. Jerome H. Barkow, Leda Cosmides, and John Tooby (New York: Oxford University Press, 1992), 555–579; Gordon H. Orians, "An Evolutionary Perspective on Aesthetics," *Bulletin of Psychology and the Arts* 2, no. 1 (2001): 25–29.

30. David Herman, "1880–1945: Re-Minding Modernism," in *The Emergence of Mind: Representations of Consciousness in Narrative Discourse in English*, ed. David Herman (Lincoln: University of Nebraska Press, 2011), 265. David Herman's "cognitive approach" of the third wave outstrips, in my opinion, any approach that is tied more closely to "an evolved and culturally developed human nature" than it is to local ecologies. He uses an example from James Joyce's *Portrait* scene that takes place when Stephen goes to the chapel across town to confess after hearing Father Arnall's sermon about hell. "Figuring a tight coupling of intelligent agent and broader environment, Joyce's text suggests how the history of Stephen's interactions with his material and social surroundings give rise to the world that he experiences or enacts. . . . 'It was the woman: soft whispering cloudlets, soft whispering vapour whispering and vanishing.' . . . The passage in fact reflects a tension between two systems of understanding, religious and poetics/aesthetic, and suggests the extent to which Stephen oscillates between them, switching back and forth between two ways of orienting himself within his environment via two competing sets of action possibilities" (258–259). See also David Herman, *Storytelling and the Sciences of Mind*

(Cambridge, MA: MIT Press, 2013). Thanks to Brendan Shapiro for alerting me to Herman's work on this topic.

31. Ann Radcliffe, *The Romance of the Forest*, edited with an introduction and notes by Chloe Chard (Oxford: Oxford University Press, 1986).

32. William Gilpin, *Observations, Relative Chiefly to Picturesque Beauty, Made in the Year 1772, on Several Parts of England, Particularly the Mountains and Lakes of Cumberland, and Westmoreland*, Vol. 1, 2nd ed. (London: R. Blamire, 1788), 13. See also Natasha Duquette, "'The Grandeur of the Abbey': Exploring Gothic Architecture in Novels by Helen Maria Williams, Ann Radcliffe, and Jane Austen," in *Persuasions on-Line*, A Publication of the Jane Austen Society of North America, 31, no. 1 (2010). For some related criticism, see also Cynthia Sundberg Wall, *The Prose of Things: Transformations of Description in the Eighteenth Century* (Chicago: University of Chicago Press, 2006), especially chapter 8 where Radcliffe presents the reader with fully visualized settings in which events will occur.

33. Merleau-Ponty, *Phenomenology of Perception*, referencing Koffka, *Some Problems of Space Perception*: "It is certain I do have the experience of the landscape, but in this experience I'm conscious of taking up a factual situation, of bringing together a significance dispersed among phenomena, and of saying what they of their own accord mean" (263).

34. Sir Walter Scott, *The Lives of Novelists* (Philadelphia: H. C. Carey & I. Lea, 1825), 220–221.

35. Anticipating chapter 4, I give you Merleau-Ponty at just this point: "The instability of levels produces not only the intellectual experience of disorder, but the vital experience of giddiness and nausea, which is the awareness of our own contingency and the horror with which it fills us" (*Phenomenology of Perception*, 228).

Chapter Four

1. The most developed history of this model is Gail Kern Paster, *Humoring the Body: Emotions and the Shakespearean Stage* (Chicago: University of Chicago Press, 2004).

2. What is the difference between feeling and emotion? The distinction can be merely a matter of stipulated definition, though we find some real investment in whichever seems distinctly human. For instance, most would attribute feelings to a wide range of living organisms including a dog which—not who—might plausibly feel bad about chewing up the pillow but less plausibly grief stricken or remorseful. On a theoretical register Brian Massumi writes that emotions provide a "sociolinguistic fixing of the quality of an experience which is from that point onward defined as personal." *Parables for the Virtual: Movement, Affect, Sensation* (Durham: Duke University Press, 2002), 28. Contrarily, neuroscientist Antonio Damasio's influential distinction tells us "Emotions play out in the theater of the body. Feelings play out in the theater of the mind." *Looking for Spinoza: Joy, Sorrow, and the Feeling Brain* (New York: Harcourt, 1999), 28. I am interested in something else altogether, namely the emotions or feelings that are not personal, like the amusement in an amusement park. See Daniel M. Gross, *The Secret History of Emotion: From Aristotle's* Rhetoric *to Modern Brain Science* (Chicago: University of Chicago Press, 2006). For the sake of this chapter that proceeds by way of colloquial expression, I will use one of the two terms according to local usage.

3. *Wired*, http://archive.wired.com/culture/geekipedia/magazine/geekipedia/uncanny_valley. "This is the uncanny valley, the point at which an automaton looks too human for comfort. The yuck factor arises from subverted expectations, says roboticist Masahiro Mori, who coined the phrase in the 1970s." The sorites paradox provides what is for us a convenient explanation of the uncanny valley: Stimuli with human and nonhuman traits "undermine our sense of human

identity by linking qualitatively different categories, human and nonhuman, by a quantitative metric, degree of human likeness." C. H. Ramey, "The Uncanny Valley of Similarities concerning Abortion, Baldness, Heaps of Sand, and Humanlike Robots," *Proceedings of Views of the Uncanny Valley Workshop: IEEE-RAS International Conference on Humanoid Robots*: 8–13.

4. Throughout this chapter I will be quoting from Oxford World's Classics, Jane Austen, *Sense and Sensibility*, ed. James Kinsley, with an introduction by Margaret Anne Doody and notes by Clare Lamont (Oxford: Oxford University Press, 2008).

5. In fact, models of homeostasis broadly deter rhetorical analysis of emotion. In her essay, "Mixed Feelings in Aristotle's *Rhetoric*," Dorothea Frede does touch upon temporality when she explains how, following Plato in the *Philebus*, some feelings are the result of propositional attitudes that can turn out to be wrong: "cubic zirconia?!" In *Essay on Aristotle's* Rhetoric, ed. Amélie Oksenberg Rorty (Berkeley: University of California Press, 1996), 258–285, at 262. Aristotle's innovation beyond Plato, Frede argues, is the way he fine-tunes the definition of emotions around intentional objects temporally inflected ("Thanks for the lovely gift"). Inexplicably, however, Frede subsumes this innovation to a Phileban "remedial" framework where the psychic disturbance of mixed feelings resolves toward a natural equilibrium (259), and this despite the fact that Frede admits Aristotle uses the vocabulary of "mixture" nowhere at all (268). Likewise in his work on "ambivalence" following Bleuler, Sigmund Freud veers toward a homeostatic model when the therapeutic object is at issue: the individual psyche has parts or locations in more or less stable relation to one another. Thus in *Totem and Taboo* psychoanalysis is "concerned with the study of the unconscious portion [*Anteil*] of the individual psychic life" (25); while the ambivalence of human emotions can be difficult to adjust "because—there is no other way of putting it—they are localized in the subject's mind in such a manner that they cannot come up against each other" (28). So for instance a person knows nothing about his desire to touch the taboo object; at the same time he abominates that very act of touching. Quotes from *The Standard Edition of the Complete Psychological Works of Sigmund Freud*, ed. James Strachey, Vol. 13 (London: Hogart Press, 1986).

6. Gilles Fauconnier and Mark Turner, *The Way We Think: Conceptual Blending and the Mind's Hidden Complexities* (New York: Basic Books, 2002).

7. State-of-the-art hedonics are evident in these recent speeches by someone who is generally not a crass utilitarian: Ben S. Bernanke's August 6, 2012, speech to the 32nd General Conference of the International Association for Research in Income and Wealth, Cambridge, Massachusetts, available on the Federal Reserve website: http://www.federalreserve.gov/newsevents/speech/bernanke20120806a.htm; Ben S. Bernanke "The Economics of Happiness," speech delivered at the University of South Carolina commencement ceremony, Columbia, South Carolina, May 8, 2010, www.federalreserve.gov/newsevents/speech/bernanke20100508a.htm. For a welcome critique of the movement especially in its utilitarian dimension see Deirdre N. McCloskey, "Happyism: The Creepy New Economics of Pleasure," *New Republic*, June 28, 2012.

8. John Stuart Mill famously complemented Bentham with Coleridge: "The two men are each other's 'completing counterpart': the strong points of each correspond to the weak points of the other. Whoever could master the premises and combine the methods of both, would possess the entire English philosophy of their age." *London and Westminster Review* 33 (March 1840): 257–302.

9. Paul Ekman, "An Argument for Basic Emotions," *Cognition and Emotion* 6, no. 3/4 (1992): 169–200, at 170; *Emotions Revealed: Recognizing Faces and Feelings to Improve Communication and Emotional Life* (New York: Holt, 2003), 212, 69.

10. Page 213 gives us "Elinor's astonishment at this commission" from Colonel Brandon.

11. Classical French understanding linked astonishment with admiration, and astonishment

with fear. See Charles Le Brun, *A Method to Learn to Design the Passions* (Los Angeles: William Andrews Clark Memorial Library, University of California Press, 1980).

12. Surprise more often implicates a human actant—e.g., a surprise attack—as opposed to astonishment, which more often implicates non-human actants such as thunderbolt-throwing deities or in our long eighteenth century, picturesque landscapes. OED etymology compares classical Latin *attonare*: to strike with a thunderbolt, stun, stupefy. Obsolete meanings include loss of sense, or insensibility. Active meaning includes astonishment as "mental disturbance or excitement due to the sudden presentation of anything unlooked for or unaccountable," which speaks directly to our utilitarian concerns. Compare OED on surprise: "The feeling or mental state, akin to astonishment and wonder, caused by an unexpected occurrence or circumstance."

13. Ekman relabels everything beyond his seven basic emotions "affective commitments." See chapter 1.

14. "There are two very different ways to interpret a relation between an emotional incentive and the brain profile—for example, a face with a fearful expression and activity in the amygdala. The popular view attributes the relation to the induction of the state in the person that matches the symbolic meaning of the incentive; therefore, the viewer is afraid. A second, equally reasonable explanation is that the brain reacted to an event that was unexpected or unfamiliar and the viewer was surprised or experienced response uncertainty" (74). Jerome Kagan, *What Is Emotion? History, Measures, and Meanings* (New Haven: Yale University Press, 2007). For relevant history see Otniel Dror, "What Is an Excitement?" in *Science and Emotion after 1945: A Transatlantic Perspective*, ed. Frank Biess and Daniel M. Gross (Chicago: University of Chicago Press, 2014).

15. Alex Woloch, *The One vs. the Many: Minor Characters and the Space of the Protagonist in the Novel* (Princeton: Princeton University Press, 2003), 30.

16. Gilbert Ryle, "Jane Austen and the Moralists." First published in *Oxford Review*, no. 1 (1966): 5–18. Ryle is here interested in *Sense and Sensibility* and Aristotelian "equilibrium." See also Alasdair MacIntyre, *After Virtue*, 3rd ed. (Notre Dame, IN: Notre Dame University Press, 2007), 240. "A major emotion underlying Jane Austen's novels is what D. W. Harding called her 'regulated hatred' of the attitude of society to unmarried women. . . . When Jane Austen speaks of 'happiness', she does so as an Aristotelian. Gilbert Ryle believed that her Aristotelianism—which he saw as the clue to the moral temper of her novels—may have derived from a reading of Shaftesbury. C. S. Lewis with equal justice sought in her an essentially Christian writer. It is a reuniting of Christian and Aristotelian themes in a determinate social context that makes Jane Austen the last great effective imaginative voice of the tradition of thought about, and practice of, the virtues which I have tried to identify. She turns away from the competing catalogs of the virtues of the eighteenth century and restores a teleological perspective. Her heroines seek the good through seeking their own good in marriage. The restricted households of Highbury and Mansfield Park have to serve as surrogates for the Greek city-state and the medial kingdom" (240). In *The Bourgeois Virtues: Ethics for an Age of Commerce* (Chicago: University of Chicago Press, 2007), 373–376, Deirdre N. McCloskey offers Austen as an example of the seven cardinal virtues narrated for the sake of an implied audience, but no longer systematized. I would add by way of this chapter that Austen is a thorough critic of certain kinds of systematic thinking, in this case utilitarian. This essay also suggests that Austenian happiness is not Aristotelian + Christian, as MacIntyre suggests.

17. Catch trials are designed to address the Kagan surprise problem in the laboratory: "In summary, the proposed method can be used to separate the hemodynamic responses to neural events that occur in a fixed sequence within a compound trial without making any assumptions

about the shape of the hemodynamic response. The only restriction is that the experimental paradigm includes partial trials that are cognitively equivalent to the initial stage of the compound trials." J. M. Ollinger, G. L. Shulman, and M. Corbetta, "Separating Processes within a Trial in Event-Related Functional MRI: 1. The Method," *Neuroimage* 13, no. 1 (2001): 210–217, at 216.

18. Lisa Feldman Barrett, Kristen A. Lindquist, and Maria Gendron, "Language as Context for Emotional Perception," *Trends in Cognitive Science* 11, no. 8 (2007): 327–332.

19. Knight Dunlap, "Are Emotions Teleological Constructs?" *American Journal of Psychology* 44, no. 3 (1932): 572–576, at 574.

20. Instead of working exclusively in the laboratory, Paul Griffiths and Andrea Scarantino track "emotions in the wild" as they try to shift theoretical focus to neglected phenomena such as anger in a marital quarrel or embarrassment while delivering a song to an audience. See discussion in my Introduction.

21. A summary is found in Beth Innocenti, "Towards a Theory of Vivid Description as Practiced in Cicero's 'Verrine' Orations," *Rhetorica* 12, no. 4 (1994): 355–381.

22. The "Grounding Emotion" article provides this protocol in the supplementary material. "For the physical situations, the template specified the following six sentences in order: P1 described a setting and activity performed by the immersed participant in the setting, along with relevant personal attributes; S1 provided visual detail about the setting; P2A described an action (A) of the immersed participant; P2C described the consequence (C) of that action; S2 described the participant's action in response to the consequence; S3 described the participant's resulting external somatosensory experience (on the body surface). The templates for the social situations were similar, except that S1 provided auditory detail about the setting (instead of visual detail), S2 described another person' s action in response to the consequence (not action by the immersed participant), and S3 described the participant's resulting internal bodily experience (not on the body surface)." Supplementary data associated with this article can be found, in the online version, at doi:10.1016/j.neuropsychologia.2010.12.032. Here Supplementary Material p. 1.

23. Wayne Booth famously writes about Jane Austen's *Emma* in his demonstration of how a reader is invited to measure the distance between character and norm (256) in the scene, for instance, where Emma "boasts to Harriet of her indifference to marriage, at the same time unconsciously betraying her totally inadequate view of the source of human happiness" (260): "If I draw less, I shall read more; if I give up music, I shall take to carpet-work . . . I shall be very well off, with all the children of a sister I love so much, to care about. There will be enough of them, in all probability, to supply every sort of sensation that declining life can need. . . ." According to Booth, the humor of this scene "can be fully enjoyed, in fact, only by the reader who has attained to a vision of human felicity far more profound than Emma's 'comfort' and 'want' and 'need'" (261). *The Rhetoric of Fiction* (Chicago: University of Chicago Press, 1961). Though I agree with Booth and McCloskey insofar as they map how Austen's readers must be implicated in the emotional situation, I am attributing to Austen more authorial skepticism about human felicity than either McCloskey or Booth would grant. A very different and important literary-historical treatment of happiness can be found in Vivasvan Soni, *Mourning Happiness: Narrative and the Politics of Modernity* (Ithaca: Cornell University Press, 2010).

24. Published in *Organizational Behavior and Human Decision Processes* 51, no. 2 (1992): 296–312. Research psychologist Daniel Kahneman is author of the hugely popular *Thinking, Fast and Slow* and winner of the 2002 Nobel Prize in economics for his work on prospect theory, which helped launch the revolution in behavioral economics. For our purposes, prospect theory is interesting because, contra traditional utility theory, it demonstrates how relative values in any given situation matters more to an individual than absolute value calculated rationally.

Daniel Kahneman and Amos Tversky, "Prospect Theory: An Analysis of Decision under Risk," *Econometrica*, 47 (March 1979), 263–291.

25. What is utilitarian, or hedonic calculus? Bentham's *A Fragment on Government* (1776) gives us this classic formula: "It is the greatest happiness of the greatest number that is the measure of right and wrong" (Oxford: Clarendon Press, 1891), 93. So if you want to know what action is right in a given situation, Bentham would have you consider the pleasures and pains resulting from it, with respect to their *intensity, duration, certainty, propinquity, fecundity* (the chance that a pleasure is followed by other ones, a pain by further pains), *purity* (the chance that pleasure is followed by pains and vice versa), and *extent* (the number of persons affected). This is from *An Introduction to the Principles of Morals and Legislation* (Mineola, NY: Dover, 2007 [1780]), 29–30. Meanwhile this 1780 *Introduction* states, "By the principle of utility is meant that principle which approves or disapproves of every action whatsoever, according to the tendency which it appears to have to augment or diminish the happiness of the party whose interest is in question" (2). On the evolution of this principle see J. H. Burns, "Happiness and Utility: Jeremy Bentham's Equation," *Utilitas* 17, no. 1 (2005), 46–61.

26. Cf. Bentham, *Introduction*, chapter VI, "Of Circumstances Influencing Sensibility," 53: "With regard to strength of expectation; when one man expects to gain or to keep a thing which another does not, it is plain the circumstance of not having it will affect the former very differently from the latter; who, indeed, commonly will not be affected by it at all."

27. David A. Schkade and Daniel Kahneman, "Does Living in California Make People Happy? A Focusing Illusion in Judgments of Life Satisfaction," *Psychological Science* 9, no. 5 (1998): 340–346. A call for the conference "Thinking Feeling: Critical Theory, Culture, Feeling," May 2012, University of Sussex: "Happiness Is Obsolete: Uneconomic" (Theodor Adorno, *Minima Moralia*). "As the recent English riots indicate, there is no escaping the fact that economics provokes, amongst other things, strong feelings. Whether we like it or not, a neoliberal language of economics now pervades and colours our inner 'private' emotional lives; the government's emerging plans to compile a 'happiness index' is another example of how a rhetoric or discourse of 'feeling' can be co-opted by capital. More than ever, then, it is important we do not simply accept 'feeling' as a spontaneous or natural phenomenon, but instead subject it to genuinely critical scrutiny." https://centreformodernistst udiessussex.wordpress.com/thinking-feeling-conference/.

28. "Unlike Falstaff, or even Shylock, Brutus cannot simply be dismissed for attempting, in Peacham's phrase, to oppose the truth by false terms. But nor is he able to provide an unassailable justification of his act. Was he a purger or merely a murderer? It is part of his tragedy, we are made to realize, that this is a question without an answer: it will always be possible to argue in *utramque partem*, on either side of the case. Such is the power of rhetoric; more specifically, such is the power of paradiastole." In Sylvia Adamson, Gavin Alexander, and Katrin Ettenhuber, eds., *Renaissance Figures of Speech* (Cambridge: Cambridge University Press, 2007), 163.

29. Daniel Kahneman et al. describe mixed feelings in a few different ways. See for instance *Well-Being: The Foundations of Hedonic Psychology* (New York: Russell Sage Foundation, 1992) on "Objective Happiness" (9). "The bipolar nature of the Good/Bad dimension raises some difficult questions. . . . Imagine, for example, that you are out in the country during a cold night, inadequately dressed for the torrential rain, your clothes soaked. A stinging cold wind completes your misery. As you wander around you run into a large rock that provides some shelter from the fury of the wind. The event is certainly associated with a reduction in pain. Cabanac would call the experience of that moment intensely pleasurable, because he believes that the function of pleasure is to indicate the direction of a biologically significant change. However, the experience

could also be described as a composite of pleasure and pain, or perhaps as a succession of affective events in which pleasurable relief is quickly followed by a return of (diminished) distress."

30. An important and early genealogy along these lines is Albert O. Hirschman, *The Passions and Interests: Political Arguments for Capitalism before Its Triumph* (Princeton: Princeton University Press, 1977).

31. Paul Goring, *The Rhetoric of Sensibility in Eighteenth-Century Culture* (Cambridge: Cambridge University Press, 2005).

32. There has been much excellent literature on this topic, including J. G. A. Pocock's foundational work; for references and literature see Daniel M. Gross, *Secret History of Emotion*, chapter 4, "The Politics of Pride in David Hume and *David Simple.*" For a relevant reading of Jane Austen's *Pride and Prejudice* fictionalized promises in light of the 1797 Bank Restriction Act that saw prices rise as paper money proliferated, see Mary Poovey, *Genres of the Credit Economy: Mediating Value in Eighteenth- and Nineteenth-Century Britain* (Chicago: University of Chicago Press, 2008), 357–372.

33. David Hume explicitly defined causation in terms of counterfactuals when he defines a cause as "*an object followed by another, and where all the objects, similar to the first, are followed by objects similar to the second.* Or, in other words, *where, if the first object had not been, the second never had existed.*" *An Inquiry concerning Human Understanding* (Oxford: Oxford University Press, 1999 [1748]), 146. In the classic textbook *Causation in the Law*, H. L. A. Hart and Tony Honoré argue that the distinction between causes and conditions is relative to context of enquiry, or what we would call rhetoric. For example, the cause of a great famine in India may be identified by an Indian farmer as the drought, but the World Food Authority may identify the Indian government's failure to build up reserves as the cause, and the drought as a mere condition. H. L. A. Hart and Tony Honoré, *Causation in the Law* (Oxford: Clarendon Press, 1965; cited 2nd ed. 1985), 35.

34. Thanks to Jens Lloyd for alerting me to this example, Maureen Fitzsimmons for reminding me to read Kahneman, and generally to my Winter 2012 UCI graduate seminar participants for providing the rich conversation that produced this project.

35. Alex Woloch's "Character Insecurity in Austen's *Sense and Sensibility*," in *Narrative Middles: Navigating the Nineteenth-Century Novel*, ed. Caroline Levine and Mario Ortiz-Robles (Columbus: Ohio State University Press, 2011) gives us the analysis of relative value in Hegelian (not utilitarian) terms: "Even as we derive a set of consequential differences out of our varied juxtapositions of Marianne and Elinor we need also to focus on the rather unforgiving way in which *Sense and Sensibility* suggests that identity—and analysis itself—emerges only through juxtaposition. . . . An object does not possess intrinsic qualities that are directly discernible to consciousness, but only emerge as an object—with specific, defining characteristics—in negative relation to others. . . . This decline of intrinsic value is rooted in much larger socio-historical developments. Like Hegel's *Phenomenology of Mind*, a text written almost contemporaneously with Austen's work bears witness to a world in which absolute value no longer inheres within objects themselves but only emerges as the individual object or individual person is placed in radically contingent relationship to other, equally unstable, objects or persons" (29).

36. Mary Wollstonecraft, *A Vindication of the Rights of Woman* (New York: W. W. Norton, 1988), 65.

37. In an early and influential article, historian of emotion William Reddy has this to say about Paxtun marriage situations and emotional ambivalence: "The implication here is not that certain feelings such as grief at death are 'natural' or universal. Situations of intense ambivalence

are likely in most social orders because they arrange for and encourage emotional attachments or hopes that inevitably come into conflict. A young Paxtun woman is encouraged by institutional arrangements that offer both rewards and penalties to develop a deep attachment to her mother. At the same time, marriage is an essential element of the normative life cycle. Thus a wedding is highly likely to generate intense ambivalence in a bride. Intense ambivalence can be a nodal point of instability in a normative emotional style. By allowing women full expression of sorrow (but not other feelings) and by treating sorrow as a source of honor for women, Paxtun convention tries to tip the balance of ambivalence toward grief. This is not culture creating grief but convention promoting certain emotives over others because, over time, these emotives strongly influence individual emotion in a manner that allows for a certain stability and ideological comprehensibility in a community's life." William M. Reddy, "Against Constructionism: The Historical Ethnography of Emotions," in *Current Anthropology* 38, no. 3 (1997): 327–351, at 334.

Epilogue

1. *Geisteswissenschaften* is a term of the middle nineteenth century established by Schiel, Helmholtz, and Droysen as a catch-all contrary to the natural sciences, or *Naturwissenschaften.* See for instance Hans Georg Gadamer, *Truth and Method,* 2nd rev. ed., trans. Joel Weinsheimer and Donald G. Marshall (New York: Crossroad, 1989), 3–42. Anglo-American and German distinctions overlap significantly, with a tripartite classification system 1) humanities 2) human sciences 3) natural sciences complicating the transatlantic academic traditions. "Interpretive" social sciences would be grouped with humanities in a binary classification, and social science fields anchored in quantitative methodologies grouped with natural sciences.

2. Two examples among many: Patricia Clough and Jean Halley, eds., *The Affective Turn: Theorizing the Social* (Durham: Duke University Press, 2007); see also the proceedings of "The Emotional Turn in the Social Sciences," a conference held at UCLA in November of 2011. For a perspective in social science, see Margaret Weatherell, *Affect and Emotion: A New Social Science Understanding* (London: Sage, 2012), where a "situated" approach referencing Sara Ahmed and Lisa Feldman Barrett prevails.

3. Summary of mirror neuron research can be found in Marco Iacoboni, *Mirroring People: The New Science of How We Connect with Others* (New York: Farrar, Straus and Giroux, 2008).

4. Carolyn J. Dean, "Empathy, Pornography, and Suffering," *Differences: A Journal of Feminist Cultural Studies* 14, no. 1 (2003): 89–123. See her footnote 5 for a brief history of the term.

5. Andrea Scarantino and Paul Griffiths, "Don't Give Up on Basic Emotions," *Emotion Review* 3, no. 4 (2011): 444–454, at 452. More compatible with our project is scientific psychologist Jerome Kagan's call for a moratorium on the use of single words such as "fear," urging experts to use instead full sentences to write about the emotional process. See Jerome Kagan, *What Is Emotion? History, Measures, and Meanings* (New Haven: Yale University Press, 2007), 216.

6. Steven Pinker, "Toward a Consilient Study of Literature," *Philosophy and Literature* 31, no. 1 (2007): 162–178, at 176. The problem with Pinker's approach and others like it is the vast distance between the evolutionary-psychological model and the work of literature itself. Indeed, Pinker explicitly distances the analysis from the local culture of its object: "one has to show–*independently of anything we know about the human behavior in question*—that X, by its intrinsic design, is capable of causing a reproduction-enhancing outcome in an environment like the one in which humans evolved. This analysis can't be a kind of psychology; it must be a

kind of engineering—an attempt to lay down the design specs of a system that can accomplish a goal (specifically, a subgoal of reproduction) in a particular world (specifically, the ancestral environment)" (170).

7. A sympathetic project—minus the literary humanities—is outlined by Jan Slaby, "Steps Towards a Critical Neuroscience," in *Phenomenology and Cognitive Science* 9, no. 3 (2010): 397–416. See also Suparna Choudhury and Jan Slaby, eds., *Critical Neuroscience: A Handbook of the Social and Cultural Contexts of Neuroscience* (Southgate: Wiley-Blackwell, 2012); Jordynn Jack, ed., *Neurorhetorics* (New York: Routledge, 2012).

8. Critiques of brain science of emotion include the following: Kagan cited above (2007); Alva Noë, *Out of Our Heads: Why You Are Not Your Brain, and Other Lessons from the Biology of Consciousness* (New York: Hill and Wang, 2009); Paolo Legrenzi and C. A. Umiltà, *Neuromania: On the Limits of Brain Science* (Oxford: Oxford University Press, 2011); Raymond Tallis, *Aping Mankind: Neuromania, Darwinitis and the Misrepresentation of Humanity* (Durham, England: Acumen, 2011); Ruth Leys, "The Turn to Affect: A Critique," *Critical Inquiry* 37, no. 3 (2011): 434–472; Ruth Leys, "'Both of Us Disgusted in My Insula': Mirror Neuron Theory and Emotional Empathy," in *Science and Emotions after 1945: A Transatlantic Perspective*, ed. Frank Biess and Daniel M. Gross (Chicago: University of Chicago Press, 2014), 67–95.

9. A recent summary critique of brain imaging can be found in Sally Satel and Scott O. Lilienfeld, *Brainwashed: The Seductive Appeal of Mindless Neuroscience* (New York: Basic, 2013). See also Joseph Dumit, "Critically Producing Brain Images of Mind," in *Critical Neuroscience*, ed. Suparna Choudhury and Jan Slaby, 195–225.

10. Hau Olausson et al., "Unmyelinated Tactile Afferents Signal Touch and Project to Insular Cortex," *Nature Neuroscience* 5, no. 9 (2002): 900–904.

11. Tania Singer et al., "Empathy for Pain Involves the Affective but Not Sensory Components of Pain," *Science* 303, no. 5661 (2004): 1157–1162.

12. James Swain, Pilyoung Kim, and S. Shaun Ho, "Neuroendocrinology of Parental Response to Baby Cry," *Journal of Neuroendocrinology* 23, no. 11 (2011): 1036–1041.

13. Daniel Kahneman and Amos Tversky, "Prospect Theory: An Analysis of Decision under Risk," *Econometrica: Journal of the Econometric Society* 47, no. 2 (1979): 263–291.

14. Scott Rick, Cynthia Cryder, and George Loewenstein, "Tightwads and Spendthrifts," *Journal of Consumer Research* 34, no. 6 (2008): 767–782.

15. Kathleen Vohs, Nicole Mead, and Miranda R. Goode, "The Psychological Consequences of Money," *Science* 314, no. 5802 (2006): 1154–1156.

16. Stephanie Preston and F. B. M. de Waal, "Altruism," in *The Handbook of Social Neuroscience*, ed. Jean Decety and John Cacioppo (New York: Oxford University Press: 2011), 565–585.

17. Ed Yong and Uri Simonsohn, "The Data Detective," *Nature* 487, no. 7405 (July 5, 2012): 18–19.

18. Chris D. Frith, "The Social Brain?," *Philosophical Transactions of the Royal Society B: Biological Sciences* 362, no. 1480 (2007): 671–678.

19. Simon Baron-Cohen, *Zero Degrees of Empathy: A New Theory of Human Cruelty* (London: Allen Lane, 2011).

20. Hannah Faye Chua et al., "Neural Correlates of Message Tailoring and Self-Relatedness in Smoking Cessation Programming," *Biological Psychiatry* 65 (2009): 165–168.

21. For a review of the neural theory of information processing see Michael Anderson, "Neural Reuse: A Fundamental Organizational Principle of the Brain," *Behavioral and Brain*

Sciences 33, no. 4 (2010): 245–266. For a critique of "information processing" terminology and the computational model of mind, see Alva Noë, cited above.

22. Florentin Wörgötter and Ulf Eysel, "Context, State and the Receptive Fields of Striatal Cortex Cells," *Trends in Neurosciences* 23, no. 10 (2000): 497–503.

23. Although locating an "empathy circuit" in the brain is not exactly the same thing as locating compassion primarily within the scope of personal expression, each of these explanations for pro- or anti-social behavior attribute the phenomenon to personhood fundamentally, thus mitigating against explanations that are fundamentally social: e.g., empathy requires the perception of similarity, which is socially determined and therefore can be socially transformed (see discussion below).

24. Vinod Menon, "Dissociating Prefrontal and Parietal Cortex Activation during Arithmetic Processing," *NeuroImage* 12, no. 4 (2000): 357–365.

25. Stanislas Dehaene et al., "Sources of Mathematical Thinking: Behavioral and Brain-Imaging Evidence," *Science* 284, no. 5416 (1999): 970–974.

26. Vilayanur Ramachandran and Edward Hubbard, "The Phenomenology of Synaesthesia," *Journal of Consciousness Studies* 10, no. 8 (2003): 49–57.

27. Olaf Blanke et al., "Inducing Illusory Own-Body Perceptions," *Nature* 419 (2002): 269–270.

28. E.g., Eva Bonda et al., "Neural Correlates of Mental Transformations of the Body-in-Space," *Proceedings of the National Academy of Sciences of the United States of America* 92 (1995): 11180–11184; and Gemma Calvert, Ruth Campbell, and Michael Brammer, "Evidence from Functional Magnetic Resonance Imaging of Crossmodal Binding in the Human Hetero-modal Cortex," *Current Biology* 10, no. 11 (2000): 649–657.

29. Russell Poldrack, "Can Cognitive Processes Be Inferred from Neuroimaging Data?" *Trends in Cognitive Sciences* 10, no. 2 (2006): 59–63.

30. Simon Baron-Cohen, Alan Leslie, and Uta Frith, "Does the Autistic Child Have a 'Theory of Mind'?" *Cognition* 21, no. 1 (1985): 37–46.

31. John M. Wang et al., "The Neural Bases of Acquisitiveness: Decisions to Acquire and Discard Everyday Goods Differ across Frames, Items, and Individuals," *Neuropsychologia* 50, no. 5 (2012): 939–948.

32. Antonio Damasio, *The Feeling of What Happens: Body and Emotion in the Making of Consciousness* (New York: Houghton Mifflin Harcourt, 2000), 400.

33. This historical thesis is further elaborated in Frank Biess and Daniel M. Gross, eds., *Science and Emotion after 1945*.

34. Elizabeth Phelps et al., "Performance on Indirect Measures of Race Evaluation Predicts Amygdala Activation," *Journal of Cognitive Neuroscience* 12, no. 5 (2000): 729–738. A portion of this experiment description and critique appeared in Daniel M. Gross, "Toward a Rhetoric of Cognition," *Neurorhetorics*, ed. Jordynn Jack (New York: Routledge, 2012).

35. See for example Lisa Feldman Barrett, Kristen A. Lindquist, and Maria Gendron, "Language as Context for Emotional Perception," *Trends in Cognitive Science* 11, no. 8 (2007): 327–332. Barrett's basic commitment, however, is to the "psychological construction of emotion," which she distinguishes this way: "The psychological construction research program tends to focus its scientific lens on the *within-person* operations involved in constructing emotional episodes, whereas the social constructionist research program tends to focus on the *between-person* operations." Lisa Feldman Barrett and James A. Russell, eds., *The Psychological Construction of Emotion* (New York: Guilford, 2015), 13. We would emphasize how a situated approach to emo-

tion must be able to account for relevant factors that are not just within- or between-person, but also extra-personal or beyond the person, e.g., the historical institutions of racialized hatred (racial slavery for example).

36. See for example James F. Source, Robert N. Emde, J. Campos, and Mary D. Klinnert, "Maternal Emotional Signaling: Its Effect on the Visual Cliff Behavior of One-Year-Olds," *Developmental Psychology* 21, no. 1 (1985): 195–200.

37. Ethnographic and historical considerations distinguish our approach in this essay from neurophenomenology, which integrates third-person and first-person data, or a neuro-physio-phenomenological method that integrates first, second, and third-person methods. See Francisco J. Varela, "Neurophenomenology: A Methodological Remedy for the Hard Problem," *Journal of Consciousness Studies* 3, no. 4 (1996): 330–350; Giovanna Colombetti, *The Feeling Body: Affective Science Meets the Enactive Mind*, chapter 6.

38. Alicia J. Hofelich, "The Role of Unique Personal Representations in Understanding and Responding to the Emotions of Others, in Psychology" (PhD diss., University of Michigan, 2012), 154.

39. Carolyn J. Dean, "Erasures: Writing History about Holocaust Trauma," in *Science and Emotion after 1945*, ed. Frank Biess and Daniel M. Gross, 387–414.

40. Dean continues: "The idea that we have embraced victims in this 'era of the witness' (at least Jewish victims of the Holocaust) evades to what extent this new moral economy is a part of the ongoing historical refashioning of cultural attitudes to victims. These attitudes can be understood only in relational and affective terms in which suspicion, envy, attachment, and aversion are all in the process of being reformulated in reference to a new historical context in which the Holocaust of European Jewry has become problematically the paradigmatic catastrophe. It is crucial to recognize the power of a moral economy in which victims who demonstrated no agency under pressure of circumstances were least respected, least dignified, and perhaps even suspect. The narrative of the Holocaust is one of human willfulness against the odds, of human willfulness and its power to make or unmake the world, that is always implicated in any discussion about victims and victimization, including a profound cultural investment in the moral soundness of those who suffer. The introduction of the traumatized victim of catastrophe in his or her extreme disempowerment and traumatic, deferred, and thus often empirically indecipherable grief, appears to have generated as much aversion and suspicion as sympathy" (406–407).

41. One alternative: the stable-state, dynamic-systems approach that is already common in engineering and computer science. However, the inferences one can draw from such approaches are qualitatively different than the dominant approaches, which will surely inhibit the spread of such techniques in psychology per se. E.g., see Stephanie Preston and Alicia J. Hofelich, "The Many Faces of Empathy: Parsing Empathic Phenomena through a Proximate, Dynamic-Systems View of Representing the Other in the Self," *Emotion Review* 4, no. 1 (2012): 24–33.

42. The humanities, that is to say, not just the social sciences, as one sees for instance in Joan Y. Chiao and Bobby K. Cheon, "Cultural Neuroscience as Critical Neuroscience and Practice," in *Critical Neuroscience*, ed. Suparna Choudhury and Jan Slaby, 287–303.

Index